THE
CARE AND PROPAGATION OF
ORNAMENTAL
WATERFOWL

by

J. C. LAIDLAY
of LINDORES, FIFE

BEING NOTES ON THOSE WATER-
FOWL BEST SUITED FOR THE PONDS
AND LAKES OF OUR COUNTRY

1933

British Library Cataloguing-in-Publication Data
A catalogue record for this book is available from the
British Library

Poultry Farming

Poultry farming is the raising of domesticated birds such as chickens, turkeys, ducks, and geese, for the purpose of farming meat or eggs for food. Poultry are farmed in great numbers with chickens being the most numerous. More than 50 billion chickens are raised annually as a source of food, for both their meat and their eggs. Chickens raised for eggs are usually called 'layers' while chickens raised for meat are often called 'broilers'. In total, the UK alone consumes over 29 million eggs per day

According to the Worldwatch Institute, 74% of the world's poultry meat, and 68% of eggs are produced in ways that are described as 'intensive'. One alternative to intensive poultry farming is free-range farming using much lower stocking densities. This type of farming allows chickens to roam freely for a period of the day, although they are usually confined in sheds at night to protect them from predators or kept indoors if the weather is particularly bad. In the UK, the Department for Environment, Food and Rural Affairs (Defra) states that a free-range chicken must have day-time access to open-air runs during at least half of its life. Thankfully, free-range farming of egg-laying hens is increasing its share of the market. Defra figures indicate that 45% of eggs produced in the UK throughout 2010 were free-range, 5% were produced in barn systems and 50% from cages. This compares with 41% being free-range in 2009.

Despite this increase, unfortunately most birds are still reared and bred in 'intensive' conditions. Commercial hens usually begin laying eggs at 16–20 weeks of age, although production gradually declines soon after from approximately 25 weeks of age. This means that in many countries, by approximately 72 weeks of age, flocks are considered economically unviable and are slaughtered after approximately 12 months of egg production. This is despite the fact that chickens will naturally live for 6 or more years. In some countries, hens are 'force molted' to re-invigorate egg-laying. This practice is performed on a large commercial scale by artificially provoking a complete flock of hens to molt simultaneously. This is usually achieved by withdrawal of feed for 7-14 days which has the effect of allowing the hen's reproductive tracts to regress and rejuvenate. After a molt, the hen's production rate usually peaks slightly below the previous peak rate and egg quality is improved. In the UK, the Department for Environment, Food and Rural Affairs states 'In no circumstances may birds be induced to moult by withholding feed and water.' Sadly, this is not the case in all countries however.

Other practices in chicken farming include 'beak trimming', this involves cutting the hen's beak when they are born, to reduce the damaging effects of aggression, feather pecking and cannibalism. Scientific studies have shown that such practices are likely to cause both acute and chronic pain though, as the beak is a complex, functional organ with an extensive nervous supply. Behavioural evidence of pain after beak trimming in layer hen chicks has been based on the observed reduction in pecking behaviour, reduced activity and social behaviour, and increased sleep duration. Modern egg laying breeds also frequently suffer from osteoporosis which results in the chicken's skeletal system being weakened. During egg production, large amounts of calcium are transferred from bones to create egg-shell. Although dietary calcium levels are adequate, absorption of dietary calcium is not always sufficient, given the intensity of production, to fully replenish bone calcium. This can lead to increases in bone breakages, particularly when the hens are being removed from cages at the end of laying.

The majority of hens in many countries are reared in battery cages, although the European Union Council Directive 1999/74/EC has banned the conventional battery cage in EU states from January 2012. These are small cages, usually made of metal in modern systems, housing 3 to 8 hens. The walls are made of either solid metal or mesh, and the floor is sloped wire mesh to allow the faeces to drop through and eggs to roll onto an egg-collecting conveyor belt. Water is usually provided by overhead nipple systems, and food in a trough along the front of the cage replenished at regular intervals by a mechanical chain. The cages are arranged in long rows as multiple tiers, often with cages back-to-back (hence the term 'battery cage'). Within a single shed, there may be several floors contain battery cages meaning that a single shed may contain many tens of thousands of hens. In response to tightened legislation, development of prototype commercial furnished cage systems began in the 1980s. Furnished cages, sometimes called 'enriched' or 'modified' cages, are cages for egg laying hens which have been designed to overcome some of the welfare concerns of battery cages whilst retaining their economic and husbandry advantages, and also provide some of the welfare advantages of non-cage systems.

Many design features of furnished cages have been incorporated because research in animal welfare science has shown them to be of benefit to the hens. In the UK, the Defra 'Code for the Welfare of Laying Hens' states furnished cages should provide at least 750 cm^2 of cage area per hen, 600 cm^2 of which should be usable; the height of the cage other than that above the usable area should be at least 20 cm at every point and no cage should have a total area that is less than 2000 cm^2. In addition, furnished cages should provide a nest, litter such that pecking and scratching are possible, appropriate perches allowing at least 15 cm per hen, a claw-shortening device, and a feed trough which may be used without restriction providing 12 cm per hen. The practice of chicken farming continues to be a much debated area, and it is hoped that in this increasingly globalised and environmentally aware age, the inhumane side of chicken farming will cease. There are many thousands of chicken farms (and individual keepers) that treat their chickens with the requisite care and attention, and thankfully, these numbers are increasing.

CEREOPSIS GOOSE.

PREFACE

IT is not necessary to possess great lakes or even very large ponds in order to keep ornamental waterfowl successfully. Quite small ponds may be made very attractive by the addition of a pair of Mandarin or Carolina Ducks, and one has often regretted the wasted opportunities of those possessing nice ponds, by their failure to furnish them with a pair or two of ornamental Waterfowl or Flamingoes. Waterfowl have always been great favourites of mine, and I have envied those who have the facilities to keep these birds under the best conditions, for they are amongst the most satisfactory subjects for aviculture, being hardy and very beautiful and presenting just a sufficient amount of difficulty in the rearing of their progeny to make success very satisfying.

When Mr Laidlay told me that the idea of writing a book on the subject had crossed his mind, I did all in my power to encourage him to do so, for I know of no one more suited to the task. Not only has he ideal conditions under which to keep and breed his Waterfowl, but he has made the very best use of his opportunities, and given a great deal of his life, to his chosen subject.

Moreover, he does not hesitate to give away the many secrets that have been revealed to him through a long period of failures and successes.

And the result is an extremely useful volume which I much hope will lead to an increase in the number of Waterfowl keepers.

DAVID SETH-SMITH.

FOREWORD

THERE are many Wild Fowl, but not all can be kept easily, or in health, in captivity ; many also are difficult to obtain. Therefore, I have dealt mainly with those that are most suitable for ornamental waters, and tried to show how easy they are to manage when adult, to rear when young, how inexpensive to feed, and throughout a perpetual delight to watch. The illustrations are all taken from living specimens, and I am most grateful to Flight-Lieut. GODFREY and Mr J. BERRY for all the time and patience they have devoted here. Also, I am deeply indebted to Lord GREY OF FALLODEN and Mr A. EZRA. And, lastly but not least, to Mr D. SETH-SMITH for all his help and advice.

J. C. LAIDLAY.

LINDORES,
FIFE, SCOTLAND.

INDEX

8

INDEX

INDEX

PAIR OF RED FLAMINGOES

FLAMINGO

PHOENICOPTERUS ROSEUS

NATIVE of Spain, Southwards across Africa. Sexes are alike, the general colour being white just tinged with pink, flight feathers black, rest of wing reddish pink, legs and bill pink, with black tip to bill. Young in the first year have no pink. Although never bred in this country they are fairly often imported and are certainly most ornamental birds. They require to be fed on wheat and shrimps—if they will take a little milk, it is good for them. They have the name of not being hardy, but I have known one which is still alive and well living now in its sixth year within a few miles of Edinburgh in an open aviary, one end is covered in but a pond is at the open end, which the bird never leaves. The difficulty in keeping them is that they sleep standing in the water, which if it freezes, freezes in their legs. If kept on waters which did not freeze or driven into a shed during hard weather they do well. Also if they can get a certain amount of natural food they do not require the shrimps. They are very gregarious amongst themselves and quite harmless to other waterfowl. The two other species, the Red and the Argentine are also imported, but are more delicate than the Common and the colour of the Red fades often in captivity.

SWANS

To be seen at their best, swans should not be mixed with geese or duck, as they outsize them to the extent of making the swans look clumsy, nor is it safe to keep

PAIR OF ICELANDIC WHOOPERS

many of the swans with smaller waterfowl, as they will kill the young birds. To be seen at their best and kept in good condition, swans require large waters with an abundance of water weeds at a depth they can easily reach, and tall reed beds and small islands for nesting. Swans are rather slow to pair, but once mated remain so for life. They are hardy birds and easy to feed even if water plants are lacking, as they will do well on cut grass, cabbages, lettuce or such green stuff, with three or four handfuls of wheat per bird. Cygnets should be given finely chopped green food (they are very fond of duckweed), and a little biscuit meal and, when they are old enough to eat grain, this should be soaked until soft, before given them. They, like all other waterfowl, must have plenty of grit. Incubation takes about forty-two days, and entire charge of incubation and rearing of the young is best left to the birds themselves. Young should be caught up and pinioned when five or six days old, as swans if full winged are prone to wander, and the stronger or more quarrelsome ones will drive off the others.

WHOOPER SWAN
CYGNUS CYGNUS

BREEDS mainly within the Arctic Circle and sparingly in Scotland, and is the commonest of our wild swans. The average weight of an adult bird is 20 lbs., or about 6 lbs. less than that of the Mute Swan, with which swan it is sometimes confused, but is nevertheless very different, for on the water it sits with neck erect and back flat, whereas the Mute has a curved or arched carriage.   Rising from the water and flying, the wings make little sound, which applies to all wild swans, whereas the Mute's heavy beating of the air can be heard at a long distance.   White all over, with legs and feet black and bill black, with a patch of canary yellow reaching from eyes to half-way down the bill.   The call is a very clear whistling whooloop, whooloop.   The only nests I have seen were in very exposed positions, made up out of shallow water by a great quantity of small bits, only two or three inches in length, of aquatic herbage and cup shaped, with whitish eggs.   Whoopers, though decidedly ornamental and interesting birds, unfortunately are very seldom offered for sale, and then always wild-caught birds, though they have been bred in captivity, as no stock of captive-bred ones is in existence to my knowledge.   The Icelandic form of Whooper in a wild state of their own free will become very tame, and are often to be seen living in close proximity to the towns.   They are slightly smaller, and show more black on the bill than the Common Whooper does.

BEWICK SWAN
CYGNUS BEWICKII

BREEDS in Arctic Eastern Europe.   Is very like the Whooper, but average weight is about 6 lbs. less and about a foot less in length ;   white all over, with legs, feet and bill black, but the latter has a patch of canary yellow reaching from the eyes to about a quarter of the length of the bill, and not going completely over the top of the bill like in the Whooper.

The call is more of a barking whoop than the ringing note of the former bird. Although common in the winter on our coast, they are seldom caught or offered for sale in this country, but can be purchased occasionally in Holland. Young in first feather of both Whooper and Bewick are of a greyish brown, and the bills, instead of yellow, are very pale pink. The small size of this swan would make it an addition of value to any collection, and I have noticed them in a wild state feeding peacefully together with Wigeon and Wild Duck, who circle round the swans, snatching odd bits of weed brought up from a depth the duck cannot reach for themselves. So their tempers may be better than some of the other varieties.

WHISTLING SWAN
CYGNUS COLUMBRANUS

NATIVE of North America. Sexes are alike, pure white in colour, with legs, feet and bare skin on face black. The bill, which is black, shows a small patch of yellow at its base. Young in the first feather are greyish brown, with bill flesh coloured and feet pale yellow. Although fairly common in its own country, and takes readily to domestication and will withstand extreme cold better than any other swans, is seldom imported.

TRUMPETER SWAN
CYGNUS BUCCINATOR

NATIVE of America, breeding in Alaska. Sexes are alike, pure white, with bare skin on face, bill, legs and feet black, a large bird weighing up to 30 lbs.

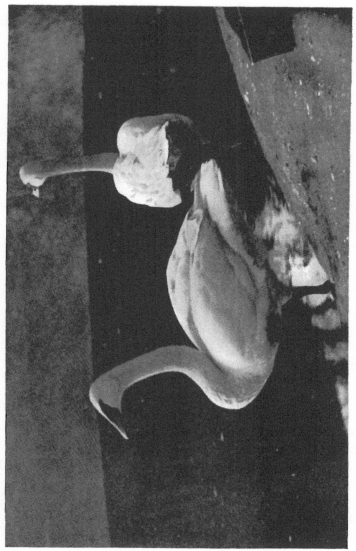

PAIR OF TRUMPETER SWANS

Young in the first feather are greyish, and the black
of the bill divided by a central flesh-coloured streak
and a patch of light bluish red on each side. Feet and
legs yellowish brown. Like most of the swans, takes
easily to domestication, but will, unless steps are
taken very rapidly, be extinct as a wild bird in its
own country. Exporting is prohibited, but an occasional specimen has been offered for sale in Europe,
where it has been bred. Mr G. H. Corsan of America
states that in all probability there are not above
thirty left in a wild state. In a book dated 1850 it
is stated that the Trumpeter Swan yielded most of
the down of commerce !

MUTE SWAN
CYNUS OLAR

ALTHOUGH wild in many parts of North and Central
Europe, is found domesticated in this country. Sexes
are alike, being all white in colour, with bare face
black, orange bill and large black knob at base,
larger in the male than in the female. Young in first
feather are light brown with slate-coloured bills
which show no knob. Graceful but most quarrelsome,
and dangerous to keep together with other waterfowl,
as they kill all young ducklings that should come near
them. And it is on record that one pair of Mutes
killed a young deer that came to the water's edge to
drink. They are birds royal, in which no subject
can have property, except by a grant from the Crown,
and are supposed to have been introduced into this
country in 1189 by Richard I.

BLACK NECKED SWAN
CYGNUS MELANCORIPHUS

Is a native of South America. Has neck of a deep blackish brown, otherwise pure white. The bill is plumbeous with a rose-coloured knob at base, larger in the male. At first glance one might imagine so strikingly marked birds would run little chance when wild, yet few have better protective colour, for the white of the body in bright sunlight melts into the water, and the dark neck is lost in the shadows of reeds or banks they may be near. Unfortunately, like the Black Swan, they are not too good tempered, and will occasionally kill young duck. They are fairly hardy and easy to keep, but are early layers, and I have known them hatched by first week in April. They hatch with yellow down, which in a couple of weeks changes to white and tan. Young in first feather show some pale brown and necks of a lighter tone than the adults. By nature they are extremely aquatic, rarely leaving the water. They are seldom for sale, as the stock is limited. Those bred in Europe are much hardier than the imported birds, more especially so after they have passed their first winter. They do better on shallow ponds than deep ones.

BLACK SWAN
CHENOPIS ATRATA

NATIVE of Australia, as its name denotes, is entirely black with white flight feathers and bill red, crossed by a bar of white. The inner feathers of the wings are curled and raised. The young in first feather are a

PAIR OF BLACK-NECKED SWANS AND YOUNG

AUSTRALIAN BLACK SWAN AND YOUNG

greyish brown. An extremely hardy and prolific species, often rearing six young at a time, but unfortunately often of a savage disposition to other swans, being able to master even the Mute. They are commonly kept and, though ornamental, lack to my mind the grace and dignity of either the Whooper or Bewick. They have been known to rear two broods in the year, a most unusual proceeding for any waterfowl. Though they come from a warm climate, they can stand extreme cold, and have lived out of doors in perfect health in America when the temperature registered twenty-four degrees below zero.

COSCOROBA SWAN
COSOROBA COSCOROBA

NATIVE of South America. This is the smallest of the swans and, unlike the rest, has no bare skin on the face. It is generally more goose-like, and comes out of the water on to the land more than do any of the other swans. Pure white in colour, with bill and feet pale red. They are less inclined to be bad tempered than the other swans, and I have seen them kept on several occasions among the smallest waterfowl. They are rarely imported or sold, and sexes are very similar, making it difficult to be sure of purchasing a pair. They are fairly hardy, and have lived in the South unprotected for long periods. They have been bred in captivity. Most swans are long lived, and there is record of a Mute living to the great age of one hundred and two.

PAIR OF COSCOROBA SWANS

SEMI-PALMATED GOOSE

SEMI-PALMATED or PIED GOOSE
ANSERANAS SEMI PALMATA

NATIVE of Australia. Sexes are alike ; head, neck, wings and tail black, rest white. The feet are only half webbed, and the bill is duck-like, so are their habits of feeding. The sex difference is noticeable in the voice, the female being less harsh than the male. Young in first feather are like the adults, but show some black on their backs. The nest is placed in reed beds, and the eggs are white. There is no eclipse, and when moulting, unlike other waterfowl, the flight feathers are gradually shed, thus retaining the power of flight, like in land birds. Before the war they were fairly frequently imported, but now rarely so.

SPUR-WINGED GOOSE
PLECTROPTERUS GAMBENSIS

NATIVE of Gambia to the Zambesi. Sexes are alike in plumage, but the female is about half the size of the male and is less bare on the face, and lacks a certain amount of the drake's gloss. The adult male has face to eyes bare and red, with flesh-coloured bill and legs. Throat and underparts white, back a glossy greeny black. Young in first feather are of a dull brown. The nest is a large and clumsy structure on the ground, though the birds are fond of perching on branches. There is no eclipse. They are fairly often imported, but are quarrelsome with other waterfowl. At the same time, are hardy and easily kept, if not subjected to too long spells of hard frost. They have laid eggs in this country, but I have not heard of young being reared.

INDIAN COMB DUCK (Male)

INDIAN COMB DUCK OR
BLACK-BACKED GOOSE

SARKIDIORNIS MELANONOTA

NATIVE of Africa and India. The drake has head and neck white dotted with black spots ; the bill is black with a large crescent-shaped knob on the top of it ; back, wings and tail glossy black ; legs and feet black ; on the shoulders the black has a purple gloss ; on the wings a green and bronze gloss. The breast, which is white, is divided from the back half-way down the breast by a thin black stripe ; flanks whitish-grey. The duck is like the drake, but very much smaller, with no knob on bill and less gloss all over. Young in first feather are like the adults, but have brown where they have black. The eggs are creamy white, and the nest is in a hollow tree trunk or branch. There is no eclipse. I have never reared any young, as the birds we had were all wild caught ones. They were easy to keep and feed, and moderately hardy. They have been bred in England. They are seldom imported, and are hardly ornamental birds, as their general appearance is like that of a Muscovy. Indeed, for wild birds they look most domesticated, and are slow and clumsy in their movements.

AMERICAN COMB DUCK OR
BLACK-BACKED GOOSE

SARKIDIORNIS SYLVICOLA

NATIVE of the Argentine, is closely allied to the Indian Comb Duck, from which it differs in having the flanks black in the drake, and it is slightly smaller.

MUSCOVY

CARINA MOSCHATA

NATIVE of Mexico and Paraguay. The sexes are alike in plumage, being a glossy greeny black all over, brightest in the male, with the inner half of the wings white. The drake is nearly twice as large as the duck, and has more bare skin on the face. Bill, feet and legs are black with pink band on bill. Young in first feather show no white on the wing or bare skin on the face. Sexes can be distinguished by size and voice, as the male is practically voiceless. They nest in hollow tree trunks, and incubation takes thirty-two days. They are not often imported, no doubt due to the domestic form being common everywhere. They are hardy, but of a quarrelsome disposition.

WHITE-WINGED WOOD DUCK

ASARCORNIS SCUTULATA

NATIVE of India and Malay States. Sexes are alike, but the duck is smaller and duller in colour. The drake's head and neck are white, spotted with black ; back, wings and tail brown, glossed with green ; underparts rusty brown ; inner half of wings white ; legs yellow, feet mottled, so is bill, with base and tip black. In breeding condition, the duck develops an orange-red coloured swelling at the base of the bill. They are good natured in captivity, where they have been kept in India, but I have not heard of their being imported into this country except to the Zoological Society of London.

PINK-HEADED DUCK

PINK-HEADED DUCK
RHODONESSA CARYOPHYLLACEA

NATIVE of India. The drake's general colour is chocolate brown ; head and neck pink, with brown stripe along chin and throat ; bill pink ; legs black. The duck is of a duller edition, with brown streak on crown and a black bill. Young in first feather are brown. The nest is made in rank grass, and the eggs are pure white. The drake in eclipse takes the brown crown of the duck. They are very rarely imported into Europe, and have not bred here. They are moderately hardy, and have wintered out of doors in the South of England. In nature their diet is comprised of shellfish and waterweeds. Possibly if an abundant supply of these was obtainable, they might breed in captivity.

INDIAN GOOSE TEAL OR COTTON TEAL
NETTOPUS COROMANDELIANUS

NATIVE of India. The drake has forehead and top of head dark brown, the rest of the head and neck white, terminating in a black collar. Breast white, flanks white, finely pencilled with black. The back is dark brown, glossed with green ; bill black ; eyes red. The duck has a dark brown streak running from the bill past the eye to the ear ; the back is brown ; flanks faint brown ; and the breast dappled in brown, but lacks the black collar of the drake. Eyes are dark brown. Young in first feather resemble the duck, but young drakes show red in the eye fairly soon.

31

They are supposed to nest in hollow branches or in trunks of trees. The drake in eclipse takes the duck's brown back, but still retains some green. They have been very rarely imported, and have not bred in captivity. Apparently though fairly hardy, they are extremely bad travellers, which accounts for their so seldom being imported.

MANDARIN DUCK
AIX GALERICULATA

NATIVE of China. No other drake has the orange neck and upturned fans on each wing of the same colour, nor I think is any other drake so aware of his beauty. Mandarin drakes are continually showing off these attractions to their ducks. The duck is a general colour of plain brown above and white below, the flight feathers being silver white on the outsides, distinguishing her from all other duck except Carolina, who likewise show the same colouration in the flight feathers. She can be known from her near ally the Carolina, as the latter is darker and more glossed with green and purple on the back, and shows more white round the eye. Young in the first feather resemble the female, but are duller in colour. Young males show the red bill and orange legs, females being slatey green in the legs and bill. Eggs are laid from the middle of April and are white in colour, in size that of a rather large bantams. They nest in a hollow log or nest box made for the purpose. Several of these should be provided, more than there are pairs of duck,

MANDARIN DRAKE

as they always like some better than others. If several
lay in one box, quarrels may result, or some eggs may
be incubated while others are still being laid. Incubation is usually exactly thirty days. Males in eclipse
resemble the duck, but retain red bill and orange legs.
They breed in their first year. They are hardy,
gentle towards other waterfowl, not prone to hybridise,
and very nocturnal by nature. Because of their
fidelity to each other when paired, Mandarin are often
carried in bridal processions in China as a symbol to
the newly married couple.

CAROLINA or WOOD DUCK
AIX SPONSA

NATIVE of North America. Drakes have a large and
heavy crest, glossed green, purples and blues, with
two white stripes, one over the eye and crown, the
other from the eye to end of crest. A white and black
stripe divides the breast from the flanks—breast
maroon, flanks buff yellow. The duck resembles the
Mandarin Duck, but is darker on the back, which is
glossed with green and purple. The eye is surrounded
with white, which increases with the age of the bird.
Young in first feather are like the adult female, but
lack the gloss. Young males show a strip of white
running up from the white throat to nearly meet the
eye ; females lack this. Eggs unfortunately are often
laid very early, sometimes in the middle of March,
but the first week of April is more usual. They use
the same type of nest as Mandarin and, if kept

CAROLINA DRAKE

together, it is common to get the two species laying in the same nest. The eggs are white and slightly smaller than those of the Mandarin, though the Carolina is slightly the larger of the two duck. Incubation is from thirty to thirty-one days. The drake in eclipse resembles the duck without the white round the eye, and showing the white stripe on the cheek. They breed in their first year. Carolina are sociable and harmless to other waterfowl. Adults are very hardy, though young are somewhat difficult at first. They lay more eggs than Mandarin, but the percentage of fertility is not so high. Hybrids between them and other duck are not uncommon. In a wild state they are found during the nesting season in high timber, nests having been found at an altitude of sixty feet from the ground, from which the young, often only a few hours old, jump safely to the ground.

MANED GOOSE
CHENONETTA JUBATA

NATIVE of Australia. Although called a goose, resembles more a duck in size and colour, and in Australia is known as the Wood Duck. The male's head is blacky brown with crest ; breast is speckled grey and black ; bill, flight feathers and tail black ; legs and feet blackish grey ; wing bar green, the rest being grey finely pencilled. The female has a paler brown head with white eye streak, and the flanks mottled with white. Young birds in the first feather resemble the female. They are very gentle in disposition, and

MANED GOOSE

can be kept with waterfowl the size of Teal, but are
very seldom offered for sale. They do not breed until
their second year, and incubation lasts twenty-eight
days. In a wild state they nest in hollow trees. In
captivity they will nest in a box about 18 inches by
18 inches. The young in first feather resemble the
adults, but are slightly duller in colour.

GEESE

MOST of the geese will breed in captivity, and all can
be ranked as hardy birds, easy to keep as long as
they get plenty of short grass to graze on and have
sufficient room to breed in, for during the nesting
season they are apt to be quarrelsome amongst themselves. But if they do have plenty of room, each pair
will fix on its own nesting area and stay mainly within
that. In the autumn and winter they are gregarious,
though they will go about more or less together, the
various species keep to themselves. Sexing of the
geese is always rather a difficult matter, but as a
fairly general rule, the goose has the top of the head
rounder than that of the gander and shorter, his being
longer and flatter. In the water, especially in early
spring or when a gander is courting a goose, he swims
with his tail more cocked up, whereas the goose keeps
hers in a straight line with the body. Voice, too,
varies, the ganders in some species having a shriller
voice than the goose. If allowed to sit and rear their
own young, ganders should be fed at the nest, or they
may starve rather than leave off guard over their

YELLOW-BILLED BEAN GOOSE

sitting mate. Young geese for the first four or five weeks require only short grass, clover and buttercups to feed on, with of course plenty of grit. It is noticeable in hard frost and snow, geese with black feet are more subject to frost bite than those with flesh-coloured ones.

CEREOPSIS GOOSE
CEREOPSIS HOLLANDIAE

NATIVE of Australia, sometimes called Cape Barren Goose. Sexes are alike in colour, being a French grey all over, except for forehead, which is white, and tail and flight feathers black. On the back and wings the feathers have a black heart-shape on each ; the legs are pink and feet black ; bill is short and covered with a pea-green cere, tip black. The female is shorter in the head and more rounded in the crown than the male, and differs in voice. Young in the first feather are like the adults, but legs are grey and do not turn pink until four months old. Eggs are white, and resemble all the other geese. Nests are like other grey geese, but are placed often in unexpected sites, such as among stones or even in an open shed, with little or no attempt at concealment. Incubation takes thirty-five days. They breed in their third year. Cereopsis are very hardy and easy to rear, living entirely on grass, as long as that can be procured. They are extremely ornamental, and become strongly attached to human beings and get very tame. Unfortunately they are bad tempered with other water-

PAIR OF CEREOPSIS

fowl. However, they can be kept on lawns without water except for drinking, as fertility does not depend on a swimming depth of water and, given access to unlimited water, they only go on it to wash. In their first winter their feet are somewhat susceptible to frost bite, so an open shed should be available for the birds to go into if they wish. If snow lies on the ground they should be given some cabbage in lieu of grass, as well as some wheat.

GREYLAG GOOSE
ANSER ANSER

NATIVE of Britain, reaching South to India. The name Greylag by some is supposed to be derived from either the grey goose that lagged behind when the others migrated, or because it lived on the grey *laggio*, grey lakes, derived from Roman occupation days, but personally after examining many I have shot, and seeing the legs and feet so grey when dead, I think Greylegs to have been the origin of their name, as the bird would most often have been examined dead rather than alive. The sexes are alike, and are grey brown all over, except for the underparts and end of tail, which is white, and the shoulders of the wing French grey ; bill is orange ; feet and legs flesh colour. On the water the gander swims with his tail cocked up, the goose carrying hers in a straight line from the body. In the first year the young are duller brown, and for the first four or five months have a white blotch of feathers where the lower bill joins the throat about

PAIR OF GREYLAG

the size of a threepenny bit. Eggs are laid in the middle of April and are pure white. The nest is situated in low cover, such as heather or grass of about the same height, often at the side of a rock or low bush. Incubation takes twenty-seven days. They do not breed until three years of age. They are very hardy and easy to rear, and can be kept safely with any kind of waterfowl, however small. On several occasions when watching flocks of Wild Greylag feeding on the stubbles, I have noticed that when the goose doing sentinel wishes to be relieved, it goes up to another bird and gives it a peck, whereupon that bird takes on the duty of sentinel until it in its turn is relieved.

BEAN GOOSE
ANSER FABALIS

WINTERS in Britain and Europe and breeds in Siberia and other areas within the Arctic Circle. The darkest brown of the grey geese, with lacing to feathers less distinct, and a noticeably longer head and neck than the others. Sexes are alike. Bill black at base and tip and orange in centre ; legs orange. There is another variety or rather branch with the bill yellow and the head not quite so long or clumsy. Eggs, nest and period of incubation like that of the Greylag. Young are like the adults, only slightly darker. This is a clumsy-looking bird, and though like all the Grey Geese, harmless with other waterfowl, can hardly be classed as an ornamental species.

PAIR OF BEAN GEESE

WHITE FRONTED GOOSE
ANSER ALBIFRONS

BREEDS in the Arctic and found in winter from Britain to India and America. Sexes are alike, brown like the Greylag, but of a slightly darker shade, with pure white on forehead and side of face at bill, and lacking the French grey on the shoulders, and with much black barring underneath, where only some feathers and that only in old birds is seen on the Greylag ; bill flesh-coloured and legs dull orange. In eggs, nest and periods of incubation it is the same as the Greylag. Young in first feather show no barring underneath, but in second season some individual birds are quite heavily barred. The voice is much shriller than that of the Greylag and more musical. They are more of grass grazers than any of the other Grey Geese. They are very gentle and harmless with other waterfowl. They will cross freely with Greylag and the young are fertile.

LESSER WHITE-FRONTED GOOSE
ANSER ERYTHROPUS

NATIVE of Lapland. Sexes are alike and resemble the White-fronted Goose, but are much smaller, if anything smaller than a Brent Goose. The general colour is darker than in the larger form, the eye is encircled with yellow skin, and the voice shriller. They can be got from Holland or such places where wild fowl netting is carried on to a large extent, but only

PAIR WHITE-FRONTED GEESE

occasional ones are offered for sale, and I have heard of only one case of their breeding in captivity, though there is no reason why they should not do so. They are gentle in nature, and as easy to keep as any of the grey geese.

PINK-FOOTED GOOSE
ANSER BRACHYRHYNCHUS

BREEDS in Iceland and Spitzbergen, wintering here and the western part of the Continent. It is the rarest of the Grey Geese. Sexes are alike. The general colour is like that of the Greylag but much darker on head and neck, and the feathers on the back and wings are laced with a lighter shade than those of the Greylag's. The bill is a very distinctive mark, being black at base and tip and pinkish red in the middle ; legs and feet pink. The young in the first feather having the edges or laced parts of a duller shade than the adults, and the bill duller. Eggs are white and the nest made in the open, often on rocky slopes some distance from water. They have once been bred in confinement, namely, in Mr St Quinton's collection at Scampston Hall. The eggs unfortunately are bad travellers, as the ones we obtained from Iceland, although fresh, never hatched, or a hand-reared race might have been started. Like other geese they are great grazers, but require some grain. They can be safely kept with smaller waterfowl, and are smart and tidy in appearance.

PINK-FOOTED GOOSE

D

BAR-HEADED GOOSE
ANSER INDICUS

NATIVE of India. Sexes are alike, but the female is a little smaller. The head is white, crossed by two black bars on the top. The general colour is a light bluish grey, with head and upper part of neck white, the white running down to the bottom of the neck on each side in a narrow stripe; back brown; and front grey. The edges of the flank feathers are darker or more browny than the back, and where they meet the stern are edged with brown; tail whitey grey and primaries black; bill, legs and feet yellow. Young in first feather are brown on the head. Eggs are white and nest situated like the Grey Geese. Incubation twenty-eight days. They do not breed until their third year. Although a native of India, are quite hardy and easy to keep. They are often imported, and the majority offered for sale are wild caught birds, as the stock of hand-reared birds is small in Europe. They are good natured with smaller waterfowl.

GREATER SNOW GOOSE
CHEN HYPERBOREA NIVALIS

NATIVE of Greenland and winters on the Atlantic coast of Northern Carolina. Sexes are alike white, with black primaries, with bill, legs and feet pink. Young in the first feather show some bluish grey on head, neck, back and wings until the end of their

GREATER SNOW GOOSE

second year. Comparatively rare in captivity though they have bred, and indeed do so more readily than their near ally the Lesser Snow, from whom they can be distinguished by their greater size, being often as much as 4 lbs. heavier, by their generally clumsier and thicker build, and further by the Greater Snow often showing some greyish markings on the shoulders of the adult birds, not found on the Lesser Snows. They are, like most geese, hardy and easy to keep in captivity.

LESSER SNOW GOOSE
CHEN HYPERBOREUS

BREEDS in North Alaska. Sexes are alike, pure white except for primaries, which are black, and primary coverts bluish grey ; bill pink and gaping at the edges, which are black ; feet and legs pink with bluish tinge. Young in first feather are light greyish brown. They are pure white by November of their second year. Eggs are laid early in May and are white, nest like all other geese, incubation twenty-six days. They do not breed until their third year. They look very fine when out on grass grazing, which they are fond of doing. Easy to keep and very hardy. They are apt to hybridise in captivity with Blue Snow, and I have seen hybrids between them and Canada, the result being all white and going to the Canada in size. They can be safely kept with smaller waterfowl.

PAIR OF LESSER SNOW GEESE

BLUE SNOW GOOSE

CHEN CAERULESCENS

BREEDS in North Alaska. Sexes are alike. Head and neck are white ; back and breast dark slate, with the feathers tipped with brown ; wing coverts blue grey ; primaries black ; flanks brownish grey. Sometimes a patch of white shows just in front of legs. Bill and feet pink with bill gaping and black edged to the inside. Young in first feather have head and neck brown, and are generally of a duller shade. In the down stage they are an olive black all over and very woolly. Pure Blue Snow at this age show a white spot where the under part of the bill meets the down. Eggs are laid towards the end of April, white in colour, and take twenty-six days to hatch. Nest like that of the other geese. They do not breed until their third year. Like the Grey Geese, are hardy and easily kept, and like them are harmless with other waterfowl. They will hybridise readily with the Lesser Snow, the cross throwing almost typical Blue Snow or Lesser Snow. It is possible therefore that hybrids may occur in the wild state, as Blue Snow show some variation in markings on flanks and underparts. At the same time, I have noticed that these white parts show a tendency to increase in some birds with age, so they may be caused by age only.

BLUE SNOW GOOSE

ROSS'S SNOW GOOSE
·CHEN ROSSII

BREEDS in North Alaska. Sexes are alike. White all over except for the primaries, which are black ; legs and feet reddish pink ; bill pink, with wartlike corrugations in some individuals. This is a very small goose weighing not much over 3 lbs. Young in first feather have a tinge of browny grey over the white, and bill and feet dark grey. This is a very beautiful little bird, and unfortunately rarely imported and more rarely bred in captivity, though there are several cases of it having done so, notably in the London Zoological Gardens. It is hardy, and requires the same food and treatment as the other geese. For comparison between the four Snow Geese, the average lengths are, of the Greater Snow, 34 inches ; Lesser Snow, 25 inches ; Blue Snow, 28 inches ; and Ross's, 23 inches.

CHINESE OR SWAN GOOSE
CYGNOPSIS CYGNOIDES

NATIVE of Siberia and China. Sometimes called Swan Goose. This goose is similar to the domestic form of Chinese Goose, so well known everywhere as not to need describing, and only differing from the tame birds in having no knob at the base of bill, and being slighter and more active in build. They are hardy and noisy and seldom imported or offered for sale, because the domestic form is so common. The cross between them and Greylag is fertile, and if inbreeding is kept up with these hybrids, the Chinese Goose type dominates.

DOMESTIC FORM OF CHINESE AND YOUNG

EMPEROR GOOSE

EMPEROR GOOSE
PHILACTE CANAGICA

NATIVE of Alaska and Pacific Coast. The sexes are alike, the general colour being a blue grey with each feather barred with black, then white. Head and upper part of neck white. The abdomen is not barred, nor is the tail, which is slate, ending in white. Legs and feet orange, bill pale purple. Young in first feather are like the adults with markings less distinct, heads slate, and bill, legs and feet dusky. In a wild state the nest is very exposed, often on the open shore, and their diet is said to consist largely of shellfish and crustacea, but the few that have lived and bred in captivity are very rarely offered for sale or imported, and should require special feeding, though I have known them kept in good health when fed and treated like other geese. They have been bred in captivity, notably at Woburn Abbey in 1929.

SANDWICH ISLAND GOOSE
NESOCHEN SANDVICENSIS

NATIVE of the Sandwich Islands only. The sexes are alike. The top of head black, cheeks and neck greyish fawn as is the rest of the body, the feathers being edged with brown, flight feathers and tail black. These birds are becoming very scarce in a wild state, and very few pairs are in captivity to keep the stock going. They lay early in March, which is unfortunate, as they are delicate and require protection from frost. Taking very readily to domestication, they get very tame, and that together with the fact that they are more of a land than water bird, enables one to drive them in at night or during cold weather, so with care the race may yet be saved.

BRENT GOOSE
BRANTA BERNICLA

BREEDS in the Arctic and winters in Europe. The smallest of our geese, being about the size of a domestic or Aylesbury duck. Sexes are identical. Neck, head and breast are black, with half collar of white on neck, just below the head. The back is sooty brown, with flanks of a lighter shade and underparts in some white others dark. These are divided into species, but are the same in other respects, stern and rump white, tail black, as are the feet, legs and bill. Brent have never bred in captivity, which is rather extraordinary, as they are so much kept on ornamental waters and get exceedingly tame. Although not gay in colour, they are pretty and their antics amusing, as they go through much bowing and scraping to each other. No geese are more peaceable or more easily kept. Their diet consists mainly of grass. Young birds show less white on the neck than adults.

BERNACLE GOOSE
BRANTA LEUCOPSIS

BREEDS within the Arctic Circle, wintering in Western Europe. Sexes are alike, but some females show a darker marking from bill to eye than do the males.

BRENT GEESE

The face is cream and so is forehead ; the top of head, neck and breast black ; flanks light bluish grey faintly mottled with brown ; back slate, with feathers laced with white and black ; tail and tips of wings black, and so are feet, legs and bill. Eggs are white and smaller than those of the Grey Geese. In captivity the nest is much like that of the other geese ; incubation twenty-eight days. In the first feather the young have the black on back and wings brownish. A most gentle bird with others and very ornamental. They live mainly on grass.

RED BREASTED GOOSE

BRANTA RUFICOLLIS

BREEDS on the marshes between the Yenessi and Obi rivers and winters on the northern shores of the Caspian Sea. Head and neck black, with patch of chestnut on face, edged with white, and between eye and bill a small patch of white ; breast chestnut edged with black then white ; flanks black edged with white and underparts black to legs and from legs back to tail white ; tail black ; back black ; wings black ; feet, legs and bill black. Sexes are alike. Young birds in first feather are more of a browner shade than black, and have the chestnut parts more buff. The eggs are creamy white. This most beautiful species has so seldom been bred in captivity that authentic details of incubation are not yet to hand. In character it resembles very much the Brent. Some

PAIR OF RED-BREASTED GEESE

wild caught birds that had been kept at Woburn Abbey bred fifteen years after their arrival, and young were reared.

CANADIAN GOOSE
BRANTA CANADENSIS

NATIVE of North America. Sexes are alike ; head and neck black with white patch on face behind eye running right round to other eye. The breast is very light greyish brown ; back dark brown with feather edged with lighter brown ; flanks and underparts a lighter brown, from legs to tail white ; tail and flight feathers, black, as are legs, feet and bill. Young in first feather have feathers less clearly laced. The male is usually but not always slightly larger and coarser in the head. Eggs, which are laid in the middle of April, are white, and the nest, unlike the Grey Geese, is sometimes placed in quite thick cover. I have known them nest on thickly wooded islands. They do not breed till their third year. They look handsome on large lakes when flying about loose, in which state they can be kept, as they do not wander far and return to their home waters, if these are large enough, but they are bad tempered and bullies among smaller water-fowl. They are exceedingly common in captivity, and have little to recommend them a place among ornamental waterfowl. Canadian Geese have been found in a wild state, sitting on eggs laid in the old nests of eagles, situated in lofty trees.

GREYLAG GOOSE (FIVE DAYS OLD)

WHITE-CHEEKED GOOSE
BRANTA CANADENSIS OCCIDENTALIS

BREEDS along the Pacific Coast from Alaska to California. Whether really different from Branta Canadensis is questionable, as it resembles this goose in almost all respects, except that it shows a white collar at base of neck, *i.e.*, where the black ends and the breast is a dark grey, as opposed to the light breast of the Canada Goose. Sexes are alike. It has been bred in captivity but is rarely seen in confinement. In the down stage the young are said to be of a grass green colour, mottled with olive, whereas the young of Branta Canadensis at the same age are more of a golden olive green.

HUTCHIN'S GOOSE
BRANTA CANADENSIS HUTCHINSII

NATIVE of Arctic America. Sexes are alike and resemble the Canadian Goose in plumage, but are about 8 inches less in length, being 26 to 28 inches. They also differ in voice. Their nest, eggs and habits, in captivity at any rate, resemble those of the Canadian Goose, so the difference between the two birds is only in size. As far as appearance goes, one or other could be omitted from any collection, the Hutchins being the better bird of the two, as it is not so quarrelsome. I had once a hybrid between a Bernacle and Hutchins bred in captivity, this hybrid favouring mainly the Hutchins.

67

CACKLING GOOSE
BRANTA CANADENSIS MINIMA

NATIVE of western North America, breeding on the Yukon. The sexes are alike and resemble the White-Cheeked Goose, but has the white cheek patches, divided by a black bar, and also has more or less of a white collar at the base of the neck. In length only 24 inches as compared to the Canadian Goose's 36 inches. The surest index to the variety is the size of the bill, which resembles more that of the Bernacle than the Canada family. The young in the first feather are duller in colour and show the white collar less clearly defined. The nest is placed in rank grass or low reeds fairly near to water, and the eggs are white and about two-thirds the size of the Canadian Goose's. Incubation takes twenty-eight days. Though fairly common in captivity in its own country, where it has been bred in captivity, is not often imported into Europe.

MAGELLAN OR UPLAND GOOSE
CHLEPHAGA MAGELLANICA

NATIVE of Patagonia to Tierra del Fuego. The male has head and neck white; breast and underparts white; flanks white laced with black; back white laced black running out to grey; flight feathers and tail black, and so are legs, feet and bill. The female has head and neck buff; underparts, flanks and back buff, laced with black running out to grey laced black; flight feathers and tail black, and so is the bill, but legs canary yellow. In both sexes the

PAIR BLACK BARRED UPLAND GEESE

shoulders are white with green wing bar. Young in
first feather each resemble their own adult sexes, but
with markings less clearly defined. Eggs are laid early
in April, creamy buff in colour. The nest a mere
scrape, usually at the side of a low bush. Incubation
takes thirty days. There is no eclipse. They breed
in their third year, are hardy and easy to rear, requiring
mainly grass. They are very ornamental, but unfortunately quarrelsome, some individual ones being
worse than others. When kept among a number of
geese they usually behave all right, but are not to be
trusted with duck. Trouble is likely to begin at
the breeding season unless the waters are of a fair
size. It is rather interesting to note that we found,
when we imported some wild caught birds from
Patagonia, that they were much richer in colour and
markings and not so quarrelsome as the home bred
ones.

BLACK BARRED UPLAND GOOSE
CHLOEPHAGA INORNATA

NATIVE of Chile. Resembles the Upland Magellanica
in character and habits, but differs in plumage, in
that inornata is barred over those portions that are
pure white on the male Upland Magellanica. If the
two different types are bred together the young
mostly favour inornata, but by selection either one
or the other type can be bred true. Upland Geese can
stand considerable cold and exposure without any
sign of discomfort, but are not able to withstand long
periods of intense dry heat.

RUDDY-HEADED GOOSE

RUDDY-HEADED GOOSE
CHLOEPHAGA RUBIDICEPS

NATIVE of Falkland Islands and Tierra del Fuego. The sexes are alike, resembling the female Upland, but only half her size. Also, the markings all over are finer and more delicately pencilled. The lower parts of the legs and feet are blotched with black. Young resemble the adult, but can be sexed by voice—gander weak, goose harsh. They have been bred in captivity, but have been rarely imported.

ASHY-HEADED GOOSE
CHLOEPHAGA POLISCEPHALA

NATIVE of Patagonia. The sexes are alike, being grey on the head and neck ; chestnut breast ; flanks white, black barred ; underparts white ; back fawn. Chestnut patch on the shoulders. Wings are like Uplands, legs orange on the outside and black on the inside. Has been bred in captivity, is hardy and very ornamental, but rather rarely imported. The voice of the female resembles in a lower tone that of the Upland females. The male has a gentle twittering note. Both sexes frequently throw back their heads over their backs when uttering their call. They are great grazers of grass.

ANDEAN GOOSE
CHLOEPHAGA MELANOPTERA

NATIVE of Chile and Peru. The sexes are alike, though the male is considerably the larger. Colour rather a greyish white with black tail and flight feathers and

ASHY-HEADED GOOSE

PAIR OF ANDEAN GEESE

streaks on the back. Wing bar, metallic green. The bill is pink, tipped with black and the feet are red. Young in the first feather have brown instead of black on the back. They have bred here, but are not often offered for sale or imported. They are inclined to be quarrelsome, but easily kept as water, except for an occasional wash and to drink, is not needful, for they are more of a land than a water bird. When displaying they raise wings and tail.

TREE-DUCKS

THE Whistlers, sometimes called Whistling Tree Ducks, are a tribe to themselves, coming between the true ducks and the geese. They are quaint-looking birds standing high on their legs, rather round in the wings and short tailed. They are tree perching in habit. Their voice is a piercing whistle, alike in both sexes, as they are in colour too. They are very gregarious amongst themselves, but do not mix with other waterfowl, though they are not quarrelsome with them. Their feathering is rather soft and loose, and they are not really hardy, noticeably feeling the cold. Despite this they are frequently imported, and some varieties do stand our climate better than others. Few species have bred in captivity, despite which they are attractive and do well in aviaries that afford them some shelter in winter. They are quite easy to feed, and will do well on wheat and a little biscuit meal. When purchasing any of them, care should be taken of them for some weeks, as they are out of condition,

PAIR OF WHITE-FACED TREE-DUCK

being imported birds, and they should only be bought in the spring to give them every chance to get acclimatised. I have known well acclimatized birds, both Fulvous and Red-billed, come safely through weather during which snow fell and the temperature registered sixteen degrees of frost. They are good divers, so do best when kept on ponds of a fair depth.

WHITE-FACED TREE-DUCK
DENDROCYGNA VIDUATA

NATIVE of South Africa and South America. The face and throat are white, the rest of the head, neck and underparts black. The breast is dark chestnut ; the flanks are light greyish buff barred with dark brown ; the back is silvery buff, blotched with dark brown ; legs and feet blue grey. Young in first feather show some white on the black of the underparts. Sexes are alike. I have known this variety live for five years here in Scotland in an unheated open aviary with only a small portion covered in overhead, so they can be safely classed as fairly hardy. Mr Sibley of Wallingford, U.S.A., has successfully bred them in captivity.

FULVOUS TREE-DUCK
DENDROCYGNA FULVA

NATIVE of the same countries as the White-Faced, also including India. The head and neck dull rufous ; back black, with rufous barring ; flanks light buff with cream-coloured flecks ; bill and feet blue grey. Sexes are alike. This is the hardiest of the Whistlers, and I have known an occasional one live right through our Scottish winter out of doors. No doubt in the South it would do well. They have been bred in captivity in several countries. Eggs white in colour, incubation taking from twenty-six to twenty-eight days. Young in the first feather do not show the rufous or chestnut colour on the wing-coverts. The distribution of the Fulvous is peculiar for a non-migratory bird, as they inhabit Africa, India and Central South America.

WANDERING TREE-DUCK
DENDROCYGNA ARCUATA

NATIVE of Java eastwards to Australia. The Wandering Tree-Duck resembles the Fulvous, though slightly smaller and differs in having feet, legs and bill black, and is a little brighter in colour. In a wild state they are supposed to nest on the ground, whereas most of the Tree-Ducks, as their name implies, are tree-nesting duck. They are rather rarely imported, but seem moderately hardy. Sexes are alike. I have heard no case of their breeding in captivity.

EYTON'S TREE-DUCK
DENDROCYGNA EYTONI

NATIVE of Australia and New Zealand. The sexes are alike, being mainly brown in colour, with chestnut breast barred with black. The flank feathers are buff,

PAIR OF WANDERING TREE-DUCK

edged with black and curve up over the back. Legs and feet pink, as is the bill, but the latter is mottled with black. They have not bred in captivity and are seldom imported.

RED-BILLED TREE-DUCK
DENDROCYGNA AUTUMNALIS

NATIVE of Central America north to Texas. Side of head and top of neck grey, with yellowish forehead and top of head cinnamon ; rest of neck, breast and back cinnamon brown ; flanks grey, barred with black ; bill red and legs and feet pink. Sexes are alike. Young in first feather have flanks greyish white, with sooty bars. The nest is often high up in hollow branches, and the eggs white with greenish tinge. Incubation twenty-six to twenty-eight days. They are fairly often imported and breed freely in captivity in a warm climate but, like the rest of the Whistlers, will not stand really cold weather, though they have been bred in Northern France.

JAVA OR LESSER TREE-DUCK
DENDROCYGNA JAVANICA

NATIVE of India. Is more rarely imported than the other varieties mentioned, and is the smallest and most delicate of them. It resembles the Fulvous, but is smaller and of a lighter colour all over, the bright yellow of the eyelids make it easily recognisable and avoidable, as they cannot stand cold at all.

F

THE SHELDUCK FAMILY

ALL the Shelduck are gaudy and rather quarrelsome birds, and if space permits, look far better and are safer when kept amongst their own species. They show a similarity in having the shoulders white, speculum of greenish purple, flight and tail feathers black. They are active on their legs, and keep more to the land than do the true ducks, and will graze like geese. They nest in holes underground or clefts in rocks, occasionally in hollow limbs and trunks of trees. They do best when given some meat form of diet as well as grain. They keep, each pair, very much to themselves, and are not so gregarious as either duck or geese. Common Shelduck can often be seen with broods of thirty or more. Apparently the mother instinct is very strong, and a duck will often collect all other young, beyond her own, she can. The young, on the other hand, seem to be regardless of which duck is their mother, and will go with any one.

EGYPTIAN GOOSE

ALOPOCHEN AEGYPTIACUS

NATIVE of Africa. Sexes are alike, except that males show a dark collar round the neck and females more white round eye. Head and neck light buff, with a chocolate patch round eye; body grey buff, finely pencilled; shoulders white, with green bar on wings; bill and feet pink, flight feathers and tail black. Young in the first feather have the eyes brown,

COMMON SHELDUCK (WEEK OLD)

whereas the adults are yellow. Eggs are laid towards the middle of April, white in colour, incubation taking twenty-eight days. The nest may be in a bush or hole among rocks or such like site prepared for them. The female has a barking quack and the drake a prolonged hissing rattle. They are extremely hardy and easy to keep and rear. They are great grazers of grass like geese, but really belong to the Shelduck family. Though fairly ornamental, they are too quarrelsome to be worth keeping unless among such birds as Cereopsis or by themselves. One of the strangest friendships I have known was that of an Egyptian Goose to a sea-lion, whom she followed about and sat close beside, when the sea-lion was resting out of the water.

ORINOCO GOOSE
ALOPOCHEN JUBATUS

NATIVE of Amazona and Guiana. Sexes are identical, though the male is the larger. Head, neck and breast very light creamy buff, the tail and flight feathers black, the rest of the body a browny chestnut. The bird is very active on its legs, which are long and reddish pink, speculum green. This bird has not been bred in Britain, though it has produced hybrids with the Egyptian Goose, and is only occasionally imported. We had a number for some time, and found they did well on wheat and ate a lot of grass, but could not stand really cold weather. They are peaceful with other waterfowl, and looked decidedly ornamental on grass, where they spent most of their time, seldom

PAIR OF ORINOCO GEESE

going into the water. I have no doubt that in the South of England they would live out of doors throughout the year, but in Scotland we had to winter them indoors. The late Mr de Laveaga of California was successful in breeding a number of Orinoco.

COMMON SHELDUCK
TADORNA TADORNA

NATIVE of Europe. The drake has head and neck black, glossed with greenish purple; bill pinkish red, in the breeding season swelling to a knob on forehead; at base of neck a wide ring of white; breast chestnut, which goes in a ring over the back as well; centre of back white bordered with black; wings white with speculum metallic green and black flight feathers; underparts blackish chestnut and flanks white; legs pink. The duck is smaller and speculum is duller. Young in first feather have slate grey bills and legs and have no chestnut on breast or underparts. The young male is practically voiceless, but the duck has a quack. Eggs are laid in the first week of April, large and pure white. The nest is a mere scrape in a rabbit hole. Incubation takes twenty-eight days. The eclipse is a loss of the knob on the drake's bill. Gloss is lost all over and the chestnut band decreases in size with some dark tipped feathers showing. They do not breed until their second year, and are not in full feather till their third year. They are very hardy once over their first winter and easy to rear, but require some form of meat until a year old. They are apt to be quarrelsome in the breeding season, so if kept with other waterfowl should have plenty of room. Hybrids between them and the Ruddy Shelduck have been got on several occasions. They like to sit on grassy slopes, and come off the water more and for longer periods than do the true ducks, which remark applies to all the Shelduck family.

RADJAH SHELDUCK
TADORNA RADJAH

NATIVE of Australia. Sexes are alike, but the drake is slightly larger. Head, neck and eyes white ; breast dark chestnut ; underparts white ending in black ; back black mottled with chestnut ; wings have the usual Shelduck's speculum. Young in first feather resemble the adults, but have the top of the head and neck dull chestnut. They nest in holes in tree trunks and branches. They have been very rarely imported.

RUDDY SHELDUCK
CASARCA FERRUGINEA

NATIVE of Southern Europe and India. The drake has head a grey buff ; bill black ; legs, flight feathers and tail black, the rest of the bird a fawn chestnut ; shoulders of wings white. The duck is smaller and has white on forehead and round the eyes, otherwise like the drake. Young in the first feather are like the duck, but legs are greyish black, and they have some

87

RUDDY AND COMMON SHELDUCK (SAME AGE). Showing how former outgrows the latter

brown feathers on the wings. The young drake is voiceless and the duck has a loud call and soon shows white round eye. Eggs are laid about the middle of April and are glossy white. The nest is in a hole underground or between clefts in large boulders. Incubation takes twenty-eight days. They breed in their second year. The drake has no colour eclipse, but in late winter or early spring develops a black colour, especially so on face and head, which remains throughout the breeding season. They are hardy and easy to rear, more easily reared than the Common Shelduck, and require more green food than does the latter. The adults graze on grass like geese. They have the reputation of being quarrelsome, but personally I have not found them so, provided they are not overcrowded or kept on very small waters. They are extremely ornamental, and their call, which is often repeated, is musical and has great carrying powers. Ruddy Shelduck has been found nesting in Tibet at an altitude of 16,000 feet, probably the highest altitude any waterfowl breeds at.

GREY-HEADED or SOUTH AFRICAN SHELDUCK

CASCARA CANA

NATIVE of South Africa, resembles the Ruddy Shelduck, but in the drake has head grey and neckband brown instead of black, eyes yellow. The duck also has a grey head and shows white round the eyes, which are dark. Young in first feathers have all grey heads like the drake, but young ducks soon show white round the eyes. They are more aquatic than most of the Shelduck, and are better tempered and quite hardy, have been rarely imported, but have been bred in this country.

AUSTRALIAN SHELDUCK
CASARCA TADORNOIDES

NATIVE of Australia. The drake is to my mind one of the most handsome of all waterfowl. His head and neck are black, with white collar at base of neck. The breast is yellowish chestnut, bill, legs, feet, flight feathers and tail black, the rest of the bird being an iron grey, almost black, finely pencilled with white. The duck, who is smaller, lacks the white collar and chestnut breast, but is otherwise like the drake, except that she shows some white round the eyes. In a wild state they nest in hollow trunks of trees ; in captivity they have nested in rank grass, hatching five young, which were unfortunately killed by the drake. We found them rather subject to cold and requiring some animal food to keep them in condition, and noticed they do not graze anything like as much as the Ruddy Shelduck. The Australian Government has passed a law prohibiting their export ; a great pity, as they are most ornamental and, though export is prohibited, they are still shot in numbers in Australia—a strange travesty of law. We found them quite harmless to other waterfowl.

NEW ZEALAND SHELDUCK
CASARCA VARIEGATA

SOMETIMES called Paradise Duck, has head and neck black with green gloss, the rest being an iron grey, finely pencilled with white, except for some dull chestnut on the inner quills of the wing, legs and bill black, as are tail and flight feathers. The duck is smaller, dull chestnut pencilled with grey and pure white head and neck. Young in first feather resemble the drake, but young ducks show a little white on forehead. Eggs are laid in May and are white. The nest is sometimes in a hole underground or up in a nesting box. Incubation takes twenty-eight days. The drake has no eclipse, but the duck has, taking the drake's iron grey instead of her dull chestnut. They breed in their second year. They are ornamental but quite hopeless to keep with smaller waterfowl, being of such a savage disposition, and should really be kept by themselves. They are hardy and easy to keep and rear, but must have some animal food when young.

COMMON WILD DUCK
ANAS PLATYRHYNCHOS

SOMETIMES called Mallard. Too well known to need describing, and found in most countries of the world. Young in first feather resemble the duck, but the latter can quack and the drake is practically voiceless. The duck shows spots on the bill at the base, lacking in the drakes. Wild Duck are quarrelsome with other wild fowl. Will hybridise with all and sundry, and are not worth keeping among ornamental waterfowl. They are good mothers if plenty of food and shelter is available, and make good foster mothers for rarer species, when the eggs of such are given to them to hatch and rear. If it is decided to keep them, it is a good plan to allow two ducks to each drake, as this prevents the drakes bullying other kinds of ducks so much. Their eggs are useful for using as dummies and, when a bird has laid well and her eggs taken from her, say for example a Chiloe Wigeon, a few Wild Ducks' eggs can be given her to hatch and rear, which help to content her and make her lay another season. Duck do not seem to notice any difference in the ducklings of other species, provided they have hatched them themselves. Mr S. Shaw, writing in the year 1823, says that in the ten decoys in the vicinity of Wainfleet, upwards of thirty thousand duck were taken in the season. The best season then for catching duck was from the end of October to February. And he goes on to add that there is a parliamentary prohibition against pursuing this profitable pastime between the first of June and the first of October.

MELLER'S DUCK
ANAS MELLERI

NATIVE of Madagascar. Sexes are alike, and in colour are a sooty edition of the common wild duck. Wing bar metallic green. Young in first feather resemble adults, and can only be sexed by voice. Incubation

twenty-seven days. There is no eclipse. This bird
has only been imported within recent years, but a
stock of hand-reared birds is rapidly springing up,
though from an ornamental point of view they have
little to recommend them. They are one of the
largest of the wild duck, and are said to be excellent
eating.

AUSTRALIAN WILD DUCK
ANAS SUPERCILIOSA

NATIVE of Australia. The drake has top of head black,
face light buff with very black and clear eye strips
and another stripe running up from base of bill, bill
very dark lead colour and feet orange, speculum green.
The general colouration is a sooty brown mottled with
blackish brown, sexes are alike. The young in first
feather resemble the adult of either sex, for they are
extremely alike, voice only being the method of sexing.
Eggs are laid about the middle of May, and are a
greenish white in colour and rather dumpy in shape.
They nest in low thick cover. Incubation takes
twenty-seven days. They breed in their first year.
There is no eclipse. They are rather more nervous
and shy than other members of their family and,
though very hardy as adults, we have not found them
so hardy when young. They are peaceful among other
waterfowl, not really ornamental, though the head and
neck is quite handsome, the markings being so clearly
defined.

TWO AUSTRALIAN WILD DUCK AND FIVE CAROLINA (ALL WEEK OLD)

DUSKY DUCK

ANAS RUBRIPES

NATIVE of America. The general colouration is a dusky grey brown, the top of the head black edged with buff, neck buff, with a dark streak running from back of eye to back of neck, speculum metallic blue edged with black, bill olive, legs and feet olivaceous. Both sexes are alike, but the duck is slightly smaller and duller in tone. Young birds are like the duck, and voice the only method of sexing. Eggs are laid in the first week of April, and are a pale greenish buff. The clutch is unusually small for the Wild Duck family, usually being only six or seven. The nest is typical of the order, and incubation takes twenty-six days. There is no eclipse, and they breed in the first year. Hardy but not very prolific and rather shy, and only by its comparative rarity in this country can it be classed as an ornamental duck. In the distance they look black, and one has to be close to them to see their markings, which at the best are of dull colouration. In a wild state the young are better fitted to look after themselves than our Common Wild Duck of the same age are, for on hearing their mother's alarm note, they scatter individually and hide. Whereas our native birds swim and bunch round their mother, making themselves more of a target to vermin.

SPOTTED-BILLED DUCK
ANAS PEOCILORHYNCHA

NATIVE of India. The drake has a general colour of grey buff or stone colour mottled with blackish brown and a black stern. This rather dull colouration is well set off by pure white on the wing, shown by a long feather above the speculum, which is metallic green edged on both sides by a narrow bar of white. The bill is black in the centre, clear yellow at the tip, and has on each side at the base where the bill meets the feather an orange patch. The legs and feet are vivid orange. The duck is like the drake but is slightly smaller, and the colours of the bill not quite so bright. The young in the first feather resemble the duck, but lack yellow patch at base of bill. The young males are practically voiceless, the female's quack being the same as that of a Wild Duck's. Eggs are laid about the middle of April, are dirty grey white and rather round at both ends. They nest in any form of moderately thick cover, preferring low reeds. Incubation takes twenty-six days. There is no eclipse. They breed in the first year. .Though not really quarrelsome, will hybridise readily with any of the Wild Duck family. They are decidedly ornamental in a quiet way, and are hardy and easy to rear.

CHINESE GREY DUCK
ANAS ZONORHYNCHA

NATIVE of China and Japan. The drake is extremely like the Spotted-billed Duck, except all the markings are less clearly shown. The white on the wing is

CHINESE GREY DUCK (Drake)

G

duller and much smaller, speculum metallic blue, and the black streak from bill across the face gives it a decided sneering look. The bill lacks the two orange spots at the base so noticeable in the Spotted-Billed Duck. Sexes are alike. Young in the first feather are like the duck, and can only be distinguished by voice, as in the Spotted-Billed Duck. Eggs are laid early in May, are of a dirty white colour, and take twenty-six days to hatch. Nest is usually among rank grass or low reeds. There is no eclipse. They breed in their first year. They are very hardy and prolific, and as easy to rear as Wild Duck. On the whole, not much of an acquisition to a collection, as they give the appearance at first sight of being poorly coloured specimens of the Spotted-Billed Duck. They are good fliers and good eating, and would very readily lend themselves to naturalization.

YELLOW-BILLED DUCK
ANAS UNDULATA

NATIVE of South Africa, resembles the female of the Common Wild Duck but greyer all over in tone and more slim in build, with the bill a clear canary yellow with a black streak down the centre The speculum is metallic green. Feet and legs are black, with some dirty orange showing through. Sexes are alike, but the drake is slightly larger, and the female shows often the black in the centre of the bill broader. Young in the first feather are like the adults, and can

PAIR OF YELLOW-BILLED DUCK

only be sexed by voice. Sometimes the ducks show a spot at the base of the bill. Eggs are often laid early in March, the nest being like that of the Common Wild Duck. Eggs are large and white. Incubation takes twenty-seven days. There is no eclipse, and they breed in their first year. They are hardy and prolific and very easy to rear, good mothers if left to rear their young. Drakes are rather quarrelsome during the breeding season, and apt to hybridise with others of the same family. Despite their dull colour, the brightness of the bill makes them quite an ornamental species.

AFRICAN BLACK DUCK

ANAS SPARSA

NATIVE of Africa. Sexes are alike. General colour, drab with bars of white across the back. Speculum green, bill slate with black patch, feet orange. They are supposed to build in trees in their wild state. Sexes alike when young, but they can be distinguished by voice. Not often imported, though they have been successfully bred in Europe on several occasions. They are extremely bad tempered with other water-fowl, and if one or other of a pair should die, the surviving bird will rarely take to a new mate, as they appear to pair for life.

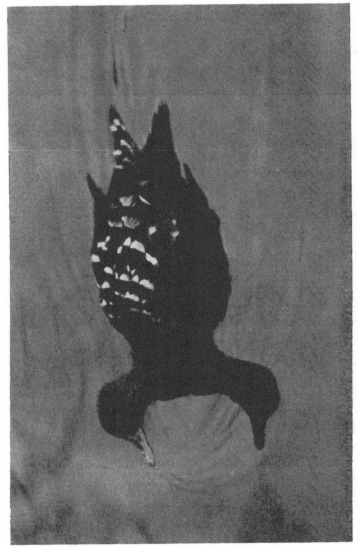

AFRICAN BLACK DUCK

BRONZE-WINGED DUCK
ANAS SPECULARIS

NATIVE of South America. Sexes are alike, face and throat white ; back, wings and tail black glossed with green, rest plain brown ; wing bar copper ; bill black ; legs orange. Little is known of this rather attractive duck, and it has seldom been imported.

CRESTED DUCK
ANAS CRESTATA

NATIVE of Peru. Sexes are alike. The head and neck are grey, general colour brown tinged with chestnut underneath. Wings have bronze bar, and tail is black and pointed. Although of a hardy and prolific family, I have not heard of their being imported but very rarely.

GADWALL
ANAS STREPERA

NATIVE of Europe, reaching South to India. The head and neck is light brown finely ticked. The breast is greyish white with each feather laced with black, the back brown. Wings have a chestnut patch on shoulder and a white bar on wing ; bill black. The duck is very like the Common Wild Duck but more finely marked, and the white bar on wing is easily noticeable. The young in first feather resemble the duck. Young males show the chestnut which is lacking in the

GADWALL DRAKE

female, and the female has one or more spots at base of bill. Both sexes quack. Eggs are laid in the first week of May, white in colour, and about the size of a Wild Duck's, slightly smaller. The nest is usually well concealed in growth such as nettles. Incubation takes twenty-seven days. Drakes in eclipse resemble the duck, except for the chestnut on shoulder and lack of spots on bill. They breed in their first year. Hardy and very prolific and easy as Wild Duck to rear, but as an ornamental bird have nothing to recommend them, except as a contrast to more brilliantly coloured ones. They are quarrelsome and very apt to hybridise with Falcated, very amorous in the spring, persecuting their own ducks to the extent of sometimes killing them. Despite their bad points, very widely kept on ornamental waters.

COMMON WIGEON

ANAS PENELOPE

BREEDS in the Arctic, and has been increasing its breeding area Southwards during the last half century. Now breeding freely in Scotland and sparingly in England. Drakes have the forehead creamy yellow, head and neck chestnut, breast magenta, general colour pencilled grey, white underneath. The duck is brown with the edges of the feathers lighter, white underneath ; bill and feet slatey blue. The young in first feather resemble the female. Young males show early the adult green speculum lacking in the female. The white shoulder of the drakes is not obtained until

the second year.   Eggs laid about 20th April or a
little earlier according to the season.   They are white,
and resemble rather those of the Carolina, but are
more pointed at the narrow end.   The nest is usually
placed in growth of moderate height, such as low
rushes, but a dry bottom to the nest is required.
Incubation takes twenty-five days.   The male in
eclipse resembles the duck, but retains the colour in
the speculum, and is of a more rich reddish hue.
They do not breed until their second year.   Common
Wigeon are very hardy and prolific, and the young
easy to rear.   They are great grazers of young grass,
and very gregarious by nature.   In the breeding
season some males become excessively amorous and
are apt to bully their ducks if these are in a backward
breeding condition.   Although Wigeon are increasing
in this country as a breeding species, their numbers
must have decreased as a winter visitor enormously,
for we read of vast catches being taken in the decoys
of a little over one hundred years ago, such as in the
decoy of Steeple in Essex, where 18,671 Wigeon were
caught in the month of October alone.

AMERICAN WIGEON
ANAS AMERICANA

NATIVE of North America.   The drake has the top
of his head a creamy white, from which he gets his
native name of Baldpate, head and neck speckled
with black on a buff cream background, breast pale
rose. The shoulders are pure white, like in the Common

Wigeon, which colour is not attained until the second year in both species. The duck is almost identical with that of the Common Wigeon duck, but can easily be known from the common species as she has the dark speculum lacking in the Common Wigeon, and the head is greyer and generally lighter in colour. Young in first feather resemble the female. Young drakes soon begin to show by their lighter coloured heads or by voice, the females having the usual purring call of all the Wigeon, young males in first feather being practically voiceless. Eggs are laid rather late, not usually before middle of May. They are white and more round at both ends than the other two varieties of Wigeon. They like to nest in rough grass. Incubation taking twenty-five days. Drakes in eclipse resemble the duck, but are a little darker. American Wigeon do not breed until their second year. They are hardy and easy to rear, but fertility is somewhat poor, doubtless due to the stock throughout Europe having got rather inbred. They should not be kept together with Common Wigeon, because of the likeness between the two ducks. By nature they are very peaceful amongst other waterfowl.

CHILOE WIGEON
ANAS SIBILATRIX

NATIVE of South America. One of the most ornamental of all the waterfowl. The drake has the face
creamy white, head and neck rather dark metallic
green, the general colour of the body being white,
laced black and flanks dull orange. Bills light slate
blue as in all three varieties of Wigeon. The duck is
like the male, but slightly smaller and duller in tone,
and does not have the bright speculum. The young
in the first feather have the black markings of a
brownier shade, and they do not show the orange on
the flanks. Young males can be distinguished from
the ducks by the colour in speculum, females lacking
it entirely. Eggs are laid in the first week of May, and
are larger than other Wigeons, white, glossy and rather
rounded at the broad end. They like to nest in fairly
thick cover, such as a low bush or rank grass. Incubation lasts twenty-five to twenty-six days. There is
no eclipse in the drake. They do not breed until
their second year. They are very affectionate birds
one to another, and keep in their true pairs throughout
the entire year, so hybrids are rare. Chiloe Wigeon
are prolific, hardy and easy to rear, requiring plenty
of grazing when young and adult. They can quite
safely be kept with any other species of waterfowl.
They do not hybridise with other duck readily, and
though we have kept them many years on the same
pond as Common Wigeon, have had only one hybrid
between the two varieties. This hybrid, a duck,
resembled very closely a Chiloe Wigeon.

HEAD OF COMMON PINTAIL DUCK

COMMON PINTAIL

ANAS ACUTA

FOUND all over the world. The drake has a long neck, the upper half of which together with the head is a dark chocolate brown with bronze lights on it ; the breast and under parts are white and the white runs up into a line through the chocolate of the lower part of the neck. General appearance—grey finely pencilled, with the tail elongated into a point, and is black at the tip. The duck resembles somewhat the Common Wild Duck, but is slighter in build and is grey where the Wild Duck looks buff. Also, the bill and feet are grey blue. Young in the first feather resemble the female, but young drakes show the cinnamon bar and bright speculum both lacking in the ducks of all ages. The latter, too, have a decided quack, young males being practically voiceless. Eggs are laid early in April, and are like those of the Common Wild Duck, but smaller and more pointed at the thin end. Pintail nest in grass, often in rather an open place. Incubation is very regular, taking twenty-three days. Drakes in eclipse do not take the ducks markings though very nearly her colour, his general look being greyer and more finely pencilled and marked than the female. They breed in their first year. Few ducks are easier to keep or rear, and are very prolific and hardy. Drakes are rather a trouble sometimes in the breeding season as they are very amorous, so much so that we found fertility the same when one drake was kept for two ducks. The hybrid between Pintail and Common Wild Duck is fertile, and the young so obtained are fertile.

CHILIAN PINTAIL

ANAS SPINICAUDA

NATIVE of South America. The drake is a grey brown mottled or laced all over with a darker shade of brown. The head is of a brighter, almost a chestnut tinge. The bill is noticeable, being yellow with a black streak down the middle ; speculum dark green. The duck resembles the drake and is only very slightly smaller. Young in the first feather resemble the adults, and lack of voice in male and quack in the female is the surest sign of sex at any age. Eggs are laid in early May, being about the size of Common Pintail, but are creamy white. Incubation takes twenty-five days. The nest, like that of the Common Pintail, is often in short grass and not too well concealed. There is no eclipse. They breed in the first year, are hardy as adults and easy to rear as young, but beyond the colour of the bill they are rather dull birds, and have not much to recommend them as ornamental water-fowl, unless to show up more gaily coloured varieties, and that they are peaceful and harmless to other birds.

BAHAMA PINTAIL

ANAS BAHAMENSIS

NATIVE of tropical South America, the West Indies and the Bahamas. The drake has cheek and throat white, general colour being fawn tinged pink and speckled with dark brown ; the bill is grey blue, with red patch on each side at the base ; tail pale buff and

BAHAMA PINTAIL.

pointed, but not so elongated as in the Common Pin-tail. The duck resembles the drake, but is smaller and the red patches on the bill paler. In first feather the young are paler all over and duller in colour than the adults but resemble them otherwise. Young ducks can be distinguished by their voice, a quack. Male practically voiceless, and by one or two black spots at base of bill, in size that of a pin head, males having none. Eggs are laid from the middle of May to first week of June, and are rich cream in colour, rather round at the broad end and large for the size of the bird. The nest is placed on dry ground in a tuft of grass or low reeds. Incubation lasts twenty-five days. There is no eclipse. I believe they do breed in their first year occasionally, but it has always been in their second year with us. Bahama are very gentle, sociable little birds, quarrelling with no others. I have never had a hybrid from them. They are hardy both as young and adult. When young the brood does best if not mixed with other varieties, unless small species such as Garganey Teal. There is no eclipse, except for an increase of colour in the red patches on the bill and an increase of gloss.

RED-BILLED DUCK
ANAS ERYTHRORHYNCHA

NATIVE of Africa. Sexes are alike. Head brown with darker crown, and cheeks white. General colour brown, mottled with a darker shade of brown ; bill red with central streak of black. It has stood our climate fairly well, but I have not heard of their breeding, and they are seldom imported.

GARGANEY TEAL

ANAS QUERQUEDULA

RANGES across Europe to India and China. The drakes have head brown, flecked finely with cream, top of head darker and a white band running from eye back ; breast brown, laced with darker brown ; flanks grey, finely pencilled, over which hang plumes, buff edged with black. The shoulders are blue grey and the back brown mottled with darker brown. The duck is grey, mottled with brown, and shoulders a dirty greyish brown. She has no colour in the speculum. Young in the first feather resemble the duck, but are easily sexed, ducks having no speculum, no blue on shoulder, spot or spots on base of bill and can quack, where young drakes are practically voiceless. Eggs are laid early in May, are white, rather long and large for the size of the bird. The nest is made in a tuft of low grass, very well concealed. Incubation lasts twenty-three days. The male in eclipse, in which state he remains longer than most other duck, resembles the duck, but can always be detected by his clear blue shoulders. They breed in their first year, are hardy, easy to rear and prolific, and though a moderately ornamental bird, have the great disadvantage of going early into eclipse and staying in it for about two months longer than any other waterfowl. They are extremely peaceful with other duck. The adult drake's voice resembles a child's rattle.

PAIR OF GARGANEY TEAL

COMMON TEAL

ANAS CRECCA

NATIVE of Europe, reaching to India and Africa.
The drake has head chestnut, slightly crested. The
eye is included in a crescent of green, edged by buff,
which reaches to end of crest ; breast pinkish brown,
feathers tipped with darker shade ; flanks grey finely
pencilled ; back grey at shoulder covers, fading off
into brown. A strip of white edged with black flanks
the back. The duck is brownish grey, with feathers
laced with a darker shade. Her speculum shows very
little colour, being mainly white, whereas the drake's
is metallic green edged with a cinnamon bar in front.
In first feather the young resemble the duck. Young
ducks quack and young drakes have practically no
voice. Females show one or more tiny black spots
at base of bill, and speculum is lacking in colour.
Eggs are laid in the middle of April, are white in colour,
and of course very small. Incubation takes twenty-one
days. The nest is a deep and cup-shaped hollow in
short grass or rather almost under it, very well hidden.
Males in eclipse resemble the duck, but lack the spots
on the bill and show the bright speculum. They breed
in their first year. They are extremely hardy as
adults and young, but are nervous duck, and though
tame enough when little, if put on to waters of any
size and having any cover, are very apt to revert to
the wild. For this reason they are hardly worth
keeping unless on very small waters. They have a
habit, too, of finding the smallest of holes in wire
netting and escaping. Yet, except for the Common
Wild Duck, no species has been longer kept in captivity. The Romans used to keep them in large
quantities.

AMERICAN GREEN-WINGED TEAL (MALE)

GREEN-WINGED TEAL
ANAS CAROLINENSIS

NATIVE of America. Drakes are identical with the Common Teal, except that they have a narrow white crescent between the termination of the breast colouration and the grey of the flank, and they lack the white strips on back. The duck is like the Common Teal Duck. They are really only a racial branch of the Common Teal, and all remarks applying to them cover both species, and unless both are kept in very small quarters where the slight differences can easily be seen, are hardly worth adding to any collection unless a represented number of species is required. Young males in the first feather are white underneath, whereas the female has brown speckling.

BLUE-WINGED TEAL
ANAS DISCORS

NATIVE of America. The drake has a blue grey head, with crescent-shaped band of white ; breast is grey brown speckled with dark brown ; flanks brown speckled darker brown and back dark brown with feathers laced with lighter brown. The shoulders are sky blue. The duck is all brown, speckled and laced with darker brown, and has blue shoulders and shows some white under throat. Young in first feather resemble the duck, but the young females lack the green in the speculum, and young males are practically voiceless. Eggs are laid in the first week of May, and are whitish buff and small, resembling those of the Common Teal in size. The nest is very well hidden in rough grass. Incubation takes twenty-three days. The drake in eclipse takes the duck's plumage, but is slightly darker. They breed in their first year, are fairly hardy and easy to keep and rear. On a small pond they show up to advantage, and are very quiet and peaceful with other waterfowl.

RING-NECKED TEAL
ANAS TORQUATA

NATIVE of Paraguay. The drakes are extremely gay little birds. The head is the colour of sandstone with the crown black, a stripe of black on the neck and half neck ring at the back of the same colour; the breast is mottled pink; flanks French grey finely pencilled; back chestnut; bill blue grey; legs and feet pink. The female is a light greyish buff, mottled with brown, with bill black. The eggs are white, and they nest in hollow branches or tree trunks, incubation taking twenty-two days. There is no eclipse. Unfortunately this attractive bird does not stand our Scottish winters well, so we have had to winter them in aviaries and eventually gave up keeping them. But they have been bred in the South of England quite successfully, and lived as far north as Northumberland throughout the hardest winter there, but did not breed. They are peaceful in disposition with other waterfowl.

PAIR OF RING-NECKED TEAL

BRAZILIAN TEAL
ANAS BRASILIENSIS

NATIVE of South America. The drake is brown, with black wings glossed with green ; the bill is red and the feet red. The duck has the bill black and head lighter brown. The young in first feather can be sexed quite easily, as young drakes show some red in the bill, and also by voice. The eggs are white and the nest placed in some cover such as a low bush. Incubation takes twenty-five days. They are hardy and easy to rear and ornamental, but not very much kept. There is no eclipse.

CHILIAN OR YELLOW-BILLED TEAL
ANAS FLAVIROSTRIS

NATIVE of South America. The drake is a very pale fawn, mottled with brown ; the bill is clear yellow, with a central black streak ; the bar on the wing is glossy black and metallic green. The duck is like the male, but slightly smaller, and the bill is not quite so bright a yellow. On the flanks she is more mottled than the drake. Young in the first feather resemble the duck, and can only be sexed by the small spot or spots on the base of the duck's bill and the voice, the female having rather a shrill quack and the young drakes being practically voiceless. Incubation takes twenty-three days, and the eggs are white and glossy, rather larger than those of the Common Teal. The nest is very well hidden in thick grass. There is no eclipse. They breed in their first year. Very peaceful

CHILIAN TEAL (MALE)

little birds, affectionate to each other, and more
ornamental than one would be led to suppose from
their description ; hardy as adults and young, and
easy to rear though a little nervous, and do better
when the broods are not mixed. In a wild state they
are supposed to be entirely a tree-nesting species,
but with us they always nest in thick grass or low
cover. They have been known to rear two broods in
the one season.

CINNAMON TEAL
ANAS CYANOPTERA

NATIVE of America. Crown of the head brownish
black, rest of head, neck, breast and flanks a rich
coppery red ; shoulders sky blue ; speculum green ;
bill black. The duck is almost exactly like that of the
Blue Winged Teal, but is a little richer in colour, and
instead of white on the throat has brownish buff.
The young in the first feather are like the duck.
Young males have underparts more streaked, whereas
the female is spotted and shows spots on bill. Eggs
are laid in the middle of May and are white in colour.
The nest is like the other Teal, well hidden and usually
in thick low grass. Incubation takes twenty-three
days. The drake in eclipse takes the duck's feather,
but is brighter in colour on the shoulders. They
breed in the first year. They are extremely gentle
little duck, and fairly hardy when adult, and the
young are not difficult to rear if given some chopped
worms and kept away from larger species until able
to fend for themselves.

FEEDING TIME

ANDAMAN TEAL
ANAS ALBOGULARIS

NATIVE of the Andaman Isles. Sexes are alike and resemble very strongly the female of the Chestnut Breasted Teal in shape and colour, but show some white round the eye. They are in all ways like the Chestnut Breasted Teal in habits of nesting and character, though more quarrelsome. They have been bred, but are very seldom offered for sale or imported, and have little to recommend them in any case.

CLUCKING TEAL
ANAS FORMOSA

NATIVE of Siberia to Japan, sometimes called Baikal Teal. The drake's head is creamy buff, black on top, with a black streak running down from the eye and a glossy green strip running back from each eye. The throat is black. Flanks are a clear slate blue, finely pencilled, over which hang very ornamental feathers striped chestnut, black and cream. The wing bar has cinnamon, metallic green, black and white in order of mentioning. The duck is a greyish brown, mottled darker brown, with a small white patch on face where the bill meets the feathers. Unfortunately this most ornamental bird has not yet been offered for sale, except as wild caught. There have been one or two recorded cases of its breeding, but no hand-reared stock has as yet been worked up. A pair bred at Fallodon in 1894 and four young were reared, but as

CLUCKING TEAL (MALE)

these were reared by the duck herself, they were to all purposes wild birds. So unfortunately,, they never bred, and their mother was killed the following year. Therefore all that come into the market, and they are often imported, are wild caught. When a number are kept together, the calling of the drakes is very amusing, being exactly like a loud and often repeated clock, clock. Drakes remain rather a long time in eclipse, during which they resemble the duck, but are darker and have not the white patch at base of bill. They are very hardy and easy to keep.

FALCATED TEAL

ANAS FALCATA

NATIVE of East Asia to Japan. The drake is a very ornamental bird, quite unmistakable owing to the prolongation of the curved feathers on the wings. These hang over the tail, almost obscuring it. The general colour is French grey, with head crested and green, glossed with bronze. Ducks are of a warm mottled brown, marked not unlike the Common Wild Duck. The bill, however, is black. The young in the first feather resemble the duck. Young drakes show a brighter speculum and more cinnamon on the bar. Young females can quack, young drakes having practically no voice. Eggs are laid about the first week of May, and are rather long shaped, about the size of a Wild Duck's and a grey cream in colour. The nest is usually well concealed in rank growth, about nettle height, on dry ground, incubation taking from twenty-six to twenty-seven days. Males in eclipse are like the ducks, but of a darker enough shade to distinguish them. They breed in their first year. Falcated are harmless to other waterfowl, very hardy as adults and easy to rear as young. I have found them very prone to hybridise with Gadwall, as the females of both species are fairly alike in appearance and voice. They will, too, occasionally cross with Chestnut Breasted Teal. They are somewhat nervous, and should not be disturbed at the nest, or are very apt to desert, though good mothers if left to rear their own young.

CHESTNUT-BREASTED TEAL

ANAS CASTANEA

NATIVE of Tasmania. The drake is a rich maroon speckled with black ; head glossy green with eyes bright red ; wing bars clearly defined and prettily marked black and bronzy green with white patch in front. In shape rather slender and long for their size. The duck is a dark brown, heavily blotched with black, and with only slightly modified shades of colour from the male in the wings and eye. Legs and feet blue grey in both sexes. In the first feather the young resemble the female, and are not very easy to sex except by voice or until the males brighter eye and richer colours begin to show, and by spots on the female's bill. Eggs are laid fairly early, and usually they start about the same time as the Carolina, in the first week of April. Eggs are about the same size as Carolina, and can be confused with them, but they are of a considerably duller or more dirty white.

Incubation takes twenty-seven days. Males in eclipse lose the majority of their colour, yet do not take entirely the duck's feather. They breed in their first year, but do better in the second year and onwards. Eggs may be laid in rank grass or in a nest put up for Carolina. In a wild state they are supposed to nest in hollow trunks and branches of trees. With us they nearly always nest in the grass. They are hardy, but not very prolific or free nesters, and are nervous. Young are easiest reared when each brood is kept separately and treated like young Carolina. They are very good-natured with other waterfowl and keep mainly to themselves.

CAPE TEAL
ANAS CAPENSIS

NATIVE of Cape Colony to Nyasaland. The sexes are alike of a fawn colour, mottled with brown, but differ from others of the Teal in that their eyes are yellow, bills pink and legs yellow. I have not heard of their being imported into this country.

COMMON SHOVELER
SPATULA CLYPEATA

FOUND all over the world. The drake has head metallic green, large shovel-shaped black bill and golden eyes, breast and feathers over shoulders white, also a patch of white between legs and stern, flanks chestnut, back brownish black. The duck is marked very like the Common Wild Duck, but her smaller size and large bill, which is horn-coloured edged at the base with dull orange, distinguish her. Young in the first feather resemble the duck. Drakes show the yellow in the eye soon and the brighter speculum, ducks of all ages having brown eyes. Eggs are laid about the middle of April, and are a greyish white sometimes with a brown tint, long and rather thin. The nest is placed in a tuft of low grass, right in the centre of it and very well concealed. Incubation takes twenty-five days. Males in eclipse take the ducks' feather, but a shade darker, and also change the colour of their bill to hers. They breed in the first year. Easy to keep when adult and hardy, but rather difficult to rear when young, requiring chopped worms for the first month, without which they do not thrive well. They should have their food given in wet state, as they like sifting amongst soft stuff for it. When newly hatched and for the first week their bills are shaped like other young duck, only slightly longer. They are very prolific if eggs are taken, often laying as many as four clutches. I have known three ducks to lay between them seventy-eight eggs in the season.

AUSTRALIAN SHOVELER

SPATULA RHYNCHOTIS

NATIVE of Australia, New Zealand and Tasmania. The drake resembles the Common Shoveler drake in eclipse, but a little richer in colour, and has head lavender, with a white crescent in front of the eye. The duck is slightly darker than the common one.

Shoveler are not good travellers. Recently the South American Shoveler has been imported, but has not yet been proved whether really hardy or not. They are decorative, the drake being somewhat like a Cinnamon Teal drake, but with head a pearl grey and bill black, the duck resembling the female of Spatula rhynchotis but with darker bill.

DIVING DUCKS

THE diving ducks, as their name denotes, get most of their food by diving, and that consists of water-snails, tadpoles, water weeds and their roots, etc. They do not do well on entirely artificial ponds unless some natural food has been introduced or is supplied in meat form. The young if being hand reared should be allowed into water after twenty-four hours for swimming, or they go back in condition. Diving duck feel the cold more than the surface feeding duck. The voice of many of the females is very much alike, being a purring growl or often repeated kurr, kurr, while the drakes emit a peculiar zumming note. When courting their ducks the eyes of Common Pochard, Rosy-billed Pochard, American Pochard, Canvasback, Red Crested Pochard, White Eyed Pochard, Tufted and Scaup intensify in colour. They almost appear to blaze with light. Those with the bills red go paler in the bill if not in good condition. Their nests are better built, deeper, and more cup-shaped than those of the surface feeders. The sexes are always different, and they are all sociable and good tempered with other waterfowl. If full winged, and a hard spell of weather

GROUP OF POCHARDS, RED CRESTED, ROSY-BILLED AND CANVAS-BACK

should come, they are very apt to leave and go to the sea coast, which is their natural movement in hard weather.

RED-CRESTED POCHARD
NETTA RUFINA

NATIVE of southern Europe and reaching to India. The drake has his crested head a rich chestnut ; neck and breast black ; back light brown ; flanks white tinged with pink in the breeding season ; bill red ; eyes vivid red. The duck is a light brown, very slightly crested on the head, which is of a darker shade, but lighter on the face and neck ; eyes dark brown ; bill blacky brown with pink tip. The young in first feather are like the duck, but young drakes soon begin to show colour in eye and bill. Eggs are laid from the end of the first week in April, and are practically identically the same as those of the Common Wild Duck. The nest is either a built up affair of aquatic weeds in shallow water, or is made on boggy land under some bent over reeds or aquatic herbage, often well concealed. Incubation takes twenty-six, sometimes twenty-seven days. The drake in eclipse resembles the duck, but retains his red bill and eyes. They breed in their first year, but give better results in second year onwards. They are very attractive duck, much given to display on the drake's part when courting. They are hardy and easy to keep and the young to rear, provided they get plenty of green food, as it is a great eater of aquatic weeds. Lawn mowings or any grass thrown on to the water is much appreciated by them.

FAVOURITE NESTING COVER FOR THE DIVING DUCK

ROSY-BILLED POCHARD (Two Weeks Old)

ROSY-BILLED POCHARD
METOPIANA PEPOSACA

NATIVE of South America. The drake has head and neck black, glossed with purple lights ; breast black, so are tail and back ; flanks light grey, finely pencilled. The bill is a vivid pink and the eye bright red. The duck is cocoa brown all over, dark eyes and bill slate. Young in first feather resemble the duck. Young drakes very early begin to show traces of pink coming on the bill, which is slate colour at first and yellow to orange in the eye, which does not get red till full plumage is acquired. Eggs are laid about the first week of May. The nest is large, well shaped, built in shallow water among high reeds. Eggs are large, rather rounded at the ends, and vary somewhat in colour, sometimes being olive green, other times more tan. Incubation takes twenty-five days. Males in eclipse have a darker grey on the flanks, and the bill takes a slightly paler hue, otherwise no change. They breed in their first year, but results are better in the second. They are very hardy both as adults and young and easy to rear, but are not very prolific, unless tall reed beds can be provided for them to nest in, that being provided they are very prolific. They seem rather a predominant species, for we had hybrids with them and Common Pochard and Red-Crested Pochard, and the resulting hybrids favoured mostly the Rosy-billed. They are peaceful with other waterfowl and are distinctly handsome birds.

WHITE-EYED POCHARD
NYROCA NYROCA

NATIVE of south Europe to India. The drake has head and neck a reddish bronze, flanks reddish brown, back brown and underparts white. The eye is dead white. The duck is a duller edition of the drake, and the eyes are a bluey grey. The young in first feather resemble the duck, but young drakes soon begin to show the white in the eyes. The eggs are white, and are laid about the middle of May. The nest is very well concealed in thick low reeds. Incubation takes twenty-four days. Males in eclipse are darker than the duck, and his gloss of colour is lost. I have not bred them before their second year, but I can see no reason why this should be, as they are in their full colour in their first year. Quite an attractive bird if one could get near them. They are unfortunately very nervous and, if kept on fair-sized ponds with reed beds, which is the kind of quarters they like, they will not often be seen. As adults they are hardy and easy to keep, but like the Tufted require some animal food, which they must have as young. They do better when reared apart from other young duck for a little time, because of their timidity. The cross between them and Tufted is fertile, and so are the young of the hybrids.

COMMON POCHARD
NYROCA FERINA

NATIVE of Europe and southwards to India. The drake has head and neck deep chestnut and very bright red eyes, breast black, body light grey finely

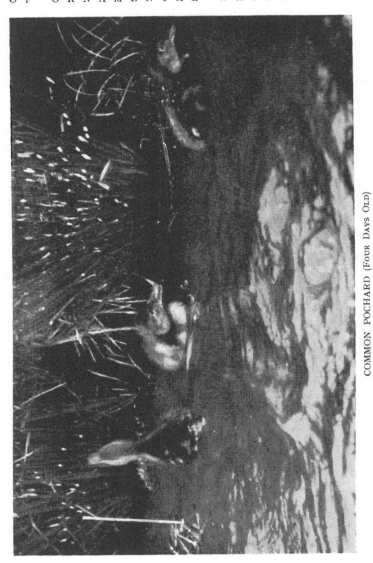

COMMON POCHARD (FOUR DAYS OLD)

pencilled. The duck is grey, but of a darker tone and blotched with brown, the head and breast brown and eyes dark brown. Bill in both sexes light slate blue and black at base and tip, foreheads receding and rather flat on top. The young in first feather resemble the female. Young drakes show soon the colour in the eye, which first looks yellowish orange, before getting the deep red of the mature drake. Eggs are laid in the first week of April. They are a little darker than those of the Wild Duck's, slightly larger and more rounded at both ends, a surprisingly large egg for the size of the bird. The nest is usually made on marshy ground in a reed bed, sometimes built up out of shallow water, and is a large clumsy affair. Incubation takes twenty-five days. The male in eclipse becomes defused with a dull wash of greyish brown over his existing colours, but does not take the female's feather. They breed in their first year. In disposition they are gentle and can be kept with any waterfowl. The young are easy to rear and the adults hardy, provided they get some water snails or such animal food. Biscuit meal with a percentage of meat in it does well. Without either they do not thrive. I have twice known hybrids between them and the Rosy-billed Pochard. These hybrids went mostly to the latter.

AMERICAN POCHARD
NYROCA AMERICANA

NATIVE of America. The drake of this variety looks like a poor specimen of the Common Pochard. That is to say, all the colours are there, but more faded and of a more browny tone. The eyes are yellow and the top of the head rounded, which distinguishes it at once. The duck is a light sooty brown all over, and has a dark brown eye. Young in first feather are like the duck. Young drakes show the yellow eye soon. Eggs are laid towards the end of April and early May, and are of a greyish white colour. The nest is built up out of shallow water in high reed beds, and often so constructed as to be often partly covered overhead by a bunch of fallen reeds. Incubation takes twenty-six to twenty-seven days. Drakes in eclipse are a dull brown. They breed in their first year. American Pochard are not easy to keep in health unless they can get a large proportion of natural food in their daily diet, nor do we find they stand a hard winter very well. They are not, however, a really ornamental bird and, as the difference between it and the Common Pochard is not great, and not in the American's favour, it is hardly worth a place in a collection that is kept purely for ornamental purposes.

CANVAS-BACK

NYROCA VALISINERIA

NATIVE of North America. The drake resembles our Common Pochard, but is much lighter in colour, and with bill long and black. The drake has head and neck brownish chestnut, breast black, tail black, body whitish grey very finely pencilled with black, eyes red, legs and feet dull grey blue. The duck has head, neck and upper part of the back brown, body grey blotched with brown, eyes brown, legs and feet same as drake, so is the bill. The young in first feather are like the duck, but young drakes soon show the colour in the eye, which is at first orange before turning red. Eggs are laid about the middle of May and are a pale grey green. Incubation takes twenty-eight days. The nest is built up out of shallow water in reed beds, and is large and clumsy. They breed best in their second year. The drake in eclipse has head of duller or more brown shade and flanks darker. They are very hardy and easy to keep if treated like the other Pochards, that is, getting some natural food. But I have found they will not nest except in reed beds. They are very peaceful with other water-fowl, though they should not be kept with Common Pochard or are sure to hybridise with them. They are decidedly ornamental and very fond of diving for water weeds, and show much activity in doing so.

BAER'S POCHARD

NYROCA BAERI

NATIVE of south-eastern Siberia. The drake has head black glossed with green, breast chocolate brown, underparts and flanks white, the rest being a dark brown. The eye is yellow. The duck has the head brown and more brown on the flanks than the drake, and her eyes are brown. The young in first feather are mostly brown of a lighter shade than the adults. They have very seldom been imported, but are quite hardy.

COMMON SCAUP

NYROCA MARILA

BREEDS in the Arctic, and in winter spreads south as far as India. The drake has head and neck black with purple and green gloss ; breast, wings and tail black ; flanks white ; back very light grey finely pencilled with black ; eyes golden ; bill and feet blue grey. The duck is all brown with a patch of white between the bill and the eye. The young in the first feather resemble the duck. The eggs are greenish-grey. Incubation takes twenty-eight days. In eclipse the drake resembles the duck. We have never been successful with this species, as we found them impossible to keep in health unless kept on waters large enough to get natural feeding for the majority of their diet, which is in nature shell-fish. Others have kept and bred them, so they can be claimed as ornamental waterfowl with justice, provided the conditions suit them. In nature their diet consists mainly of crustacea, limpets, barnacles and zostera.

TUFTED DUCK

NYROCA FULIGULA

NATIVE of Europe reaching to India. The drake's head is black glossed with purple, and is heavily crested ; eyes are golden yellow ; neck, breast, back and tail black ; flanks and underparts pure white. The duck is a dark sooty brown where the drake is black, and her white flanks are blotched with the same colour. Young in the first feather resemble the

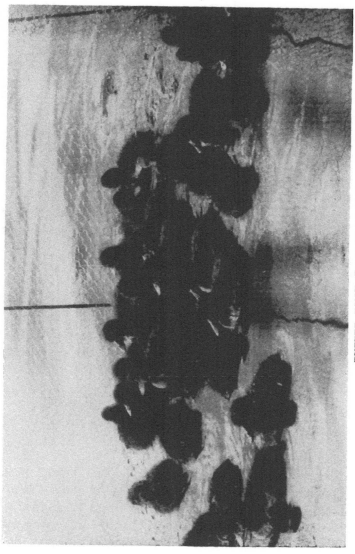

TUFTED DUCK (THREE WEEKS OLD)

duck, but young drakes feathers between and on the shoulders are very finely pencilled with grey, ducks plain brown. Some ducks in first feather show a white patch at base of bill, but not so all birds. Eggs are laid very late, often not till nearly the end of May. The site of the nest is more or less wherever the eggs can be safely laid, though they prefer boggy land with reed growths. But I have seen them on dry land in grass, in bushes, built up out of water, and even old coots' nests used. The eggs are large for the size of the bird, being as large as those of a Wild Duck's and of a brownish olive colour. Incubation lasts twenty-three days. In eclipse the drakes white flanks are pencilled and blotched with grey. They are hardy and very easy to rear, but it is impossible to keep them for any length of time in health, unless they can get some animal food, for their main diet when wild is water snails. They are very peaceable with other waterfowl. They are active and ornamental birds, and have a quaint habit of rolling over on to their backs in the water in order to preen their underneath feathers.

RING-NECKED DUCK

NYROCA COLLARIS

NATIVE of America. The drake has head, neck, breast and back black ; the head is glossed with purple and is slightly crested ; the neck is divided in the middle by a chestnut collar. Flanks pencilled black ; speculum grey ; bill black with grey-blue

PAIR OF GOLDEN-EYE

bar across ; legs and feet slate ; eyes yellow. The
duck's general colour is brown, darkest on the back,
with greyish white on the cheeks. Young in the first
feather resemble the duck. The nest is built in reed
beds. Eggs are buff coloured. Incubation takes
twenty-seven to twenty-eight days. In diet they
resemble the Scaup. Of late years they have been
hand-reared in America by the late Mr de Laveaga,
but I have not heard of their being imported into this
country as yet.

COMMON GOLDEN-EYE
BUCEPHALA CLANGULA

THE drake has slightly crested dark green head. At
the base of the bill and running nearly to the eye, is
a patch of white. Eye golden, bill grey blue. Breast
and flanks white, the latter flecked with black at the
top ; back white streaked with black ; legs orange.
The duck has a brown head, white neck and body,
and dark browny grey back. Young are like the duck.
Young drakes show a brighter eye than the females.
Eggs are laid in May and are dull green. The nest is in
a hollow branch or trunk of tree. Incubation takes
twenty-six days. The male in eclipse is like the duck,
but has more white on the shoulders and lighter yellow
eyes. They do not breed until their second year.
They are hardy, but difficult to keep without a plentiful
supply of water snails. They are very amusing to
watch when courting, as the drakes with their feet
kick up a spurt of water behind them, and go through
much bobbing and antics. They are peaceful with
other wildfowl.

BARROW'S GOLDEN-EYE
BUCEPHALA CLANGULA ISLANDICA

NATIVE of Arctic America, Iceland and Greenland. The drake has head and neck black glossed with bluish. The head is crested and has a crescentric white patch at base of the bill ; underparts white ; back black. A white patch on the wing, which is crossed by a bar of black ; flanks white, edged with black ; bill black ; legs and feet orange ; eyes yellow. The duck has head and neck brown, darkest on crown and back of neck, upper part of breast and sides grey, rest black ; the bill is brown and legs and feet orange. Young in first feather resemble the duck, but are duller. The nest is made in a hollow tree when such is available or, lacking trees, in a hole in ground. Incubation period thirty days. The eggs are of a greyish green. They have been successfully bred in captivity in this country, but to do so and to keep them in health, they require large waters abounding in water snails and shrimps. On the wing Barrow's, like the Common Golden-Eye, can be heard a long way off by the shrill whistling sound of the rapidly beating flight feathers.

LONG-TAILED DUCK
CLANGULA HYEMALIS

BREEDS in Alaska and Iceland. Head and neck white ; cheeks grey verging into chocolate ; breast brown ; flanks white ; underparts white ; wings brown, over which hang white plumed feathers. The tail is very long and pointed ; bill slatey black at base, then pink. The duck has head and neck white ; forehead and crown dusky ; face grey ; back dusky brown ; tail light brown ; underparts white. Young in first feather are like the duck. There is, however, enough difference in colour to sex the young male by the resemblance to the adult male. The Long-tailed Duck are essentially a deep-sea duck living on plankton and small shellfish. During the breeding season, when inland, they live on fresh water snails and aquatic weeds. They have been kept in captivity, but have not bred and, of course, require raw meat with the biscuit meal, and to be kept on waters where they can get some natural food. There is an eclipse.

HARLEQUIN DUCK
HISTRIONICUS HISTRIONICUS

BREEDS in the Arctic Circle and Ireland. The drake has head slatey black, with white streak running up from base of bill to crown, terminating in a chestnut streak, a small white patch behind the eye, and a streak of white down the neck and a white collar. The breast is a lighter shade than the head, terminating in a white band edged black, only reaching half-way round ; flanks chestnut ; back slatey black with white shoulder streaks ; legs, bill and feet slate. The duck resembles the Long-tailed Duck, but is a darker brown all over. Young in first feather like the duck. The nest can either be in a hole or in a hollow trunk.

147

A preference for height is shown by a pair in captivity that nested on the top of a pile of straw. Eggs are brownish grey in colour. Incubation takes thirty days. Like all deep-sea duck, they do not lend themselves readily to captivity, and they are difficult to cater for. A great pity, as they are one of the most beautiful of waterfowl. In a wild state they nest in the vicinity of fast running mountain streams, where the young live in the quickest of running waters, living on fly, insects and caddis-fly.

COMMON EIDER
SOMATERIA MOLLISSIMA

NATIVE of Northern Europe, breeding as far south as the extreme north of England. The drake has the head and upper half of the neck white, from the top of the eye and over the head glossy black, divided by white stripe in the middle, ending in apple green, of which colour there is a splash on each cheek at the ears ; breast buffy fawn ; flanks black, ending in white patch in front of tail ; rump and tail black ; back white ; bill large, long and heavy, yellowy brown on top and bluey green underneath. The female is marked not unlike the Wild Duck, but in much richer tones of brown. Young in the first year are marked like the duck, but of a more sooty colour. Sexing in the first feather is difficult, though young drakes are slightly darker than the ducks, and show a white very thin bar on wings, lacking in the females. Eggs are laid in the middle of April to end of first

COMMON EIDER AND YOUNG

week in May, are very large and dark olive green in colour. The nest is at the side of a tuft of grass in the open. Incubation takes twenty-five days if four eggs are set under a hen, but if ten are set, up to twenty-eight. Twenty-five is the normal and proper period. The male in eclipse is a dull black. They do not breed until their third year, are hardy and easy to keep if given absolutely pure water, plenty of grit, and fed on finely chopped ox liver, 2 oz. per bird per day to about five times that proportion of biscuit meal soaked and mixed together. The American Eider, somateria dresseri, which differs from s. mollissima in the bill, the frontal angles or naked portion being broader and corrugated, has been hand-reared in America.

BROODY HENS AND THEIR MANAGEMENT

TAKING it for granted that most of us who keep Waterfowl wish to rear the young, or some of them, there lies before us the choice of two methods. Either we can take the eggs, set them under hens and hand-rear the young, or we can leave the work to the birds themselves. First I will deal with the method of hand-rearing. To start at the beginning, a sufficient stock of a reliable type of broodies must be ready. After many years of trials and disappointments, during which large hens sat through whole clutches of eggs and light hens of flighty dispositions gave up sitting and ruined other clutches, we have evolved a sort of machine, a breed, the individuals of which will sit as long as is required, and are good mothers. They were evolved by breeding from pure Silkie cocks with Bantam hens of all kinds. The pullets were kept and bred the following season with Rhode Island Red and Buff Orpington cocks, resulting in birds that are larger than the usual Bantam and smaller than hens, with more and looser feathering, and possessing a most pronounced desire to go broody. They are of every colour, a point of some importance, as I will show later. When breeding from these, care has been taken to breed from those that were the best mothers and remained broody longest. These crosses seem to feel the cold more than large hens, requiring shelter from the cold winds, as the long strain of sitting and being kept in a coop naturally tells on them. So it is advisable to feed them well in the autumn and winter. Not too much freedom should be given, as they must be kept docile and easy to handle. Sleeping quarters must be clean and free of disease and vermin, or these troubles will be passed on to the young in the rearing field. When spring comes, broodies will be wanted, and it is a good plan to keep the nest-boxes in rather dark corners with plenty of hay or straw in them, and to put in each box four or five china or pot eggs, as these give them the idea of sitting.

Having got some into the required state of mind and body, broodiness is actually a sort of fever, and the birds' temperature rises considerably during this period, it is advisable to set them three or four a few days prior to that on which you actually require them, as some may be pullets or nervous, and a day or two on dummy eggs will settle them down and get them accustomed to the nest-boxes. As a matter of fact, as long as we can spare the birds to do so, we like to always have some spare broodies sitting, in case of a duck being frightened from her nest when laying, thereby causing her to desert, or for very small clutches or nests in exposed places where it would not be wise to allow the bird to complete her clutch in fear of the eggs being taken by rats, rooks or other vermin.

Nest-boxes are made to hold ten, and each is numbered and a book kept with the date of setting, number of eggs and variety or varieties in each nest, number of box and description of broody belonging to it. The point of different colours in the broodies is now plain, as otherwise describing each hen if of the same colour would be some undertaking. Broodies do best when they can be taken off and allowed loose into small wire runs, so many to each run, according to the size it is made. They can then be fed and watered and have room to dust and peck at the grass. If tied by the leg, they are frightened, and do not get the same chance to feed. When putting them back on to their eggs, the book is used, and it is easily seen that, for example, a black hen goes into number one box, a white into two, a black with yellow neck into three, and so on. After the hens have been taken out of the runs, each is moved and the ground cleaned. Inside each nest-box is shaped into a saucer-like depression, lined with short fine grass, on the bare earth. Under the earth the length and width of all the nest-boxes is buried half-inch wire netting, otherwise moles may make earthworks one night all through

SELF WITH EIDER (SEVEN WEEKS OLD)

the best sittings ; in the morning the broodies will be found standing in a sort of ploughed land with the eggs buried beneath and spoilt.

Large clutches of eggs are not as a rule a success. Nine will do better than twelve, as the heat is better distributed. There is less chance of breakages and of eggs being pushed out of the nest. Eggs vary according to species in the number of days they take to hatch, but it is quite safe to set some different varieties together that do take the same period of incubation. The young will, though of different kinds, grow up together and all be friendly one to the other. Some varieties are better not mixed, and I will make mention of these later.

There is a considerable variance of opinion as to how long a hen should be allowed off the eggs. There can be no fixed time, for it depends largely on whether the day is a hot or cold one, but in face of the fact that all duck cover their eggs with down when they leave the nest, it is obvious that eggs are better if not allowed to get really cool. I have known duck to sit three or four days on end without coming off their nests. When a duck does come off her nest she feeds and then spends some time washing and working the water all over and through her feathers, and then goes on to the nest. So nests should be kept damp, that is, the soil underneath.

After the broodies have fed and hens put back on the nests and given a short time to settle down, each should be lifted slightly up with one hand and, with the other hand, pour from a water-can with a fine rose on it tepid water over the eggs. This will soak down into the ground. When no water is visible in the nest, withdraw your hand to allow the hen to settle down. In addition to damping the nest two or three times a week, the eggs and nest should be well sprinkled with some insect powder such as Keatings. Start these operations at one end of the boxes and, having done them all, go back to where you started and take a glimpse into each to see that the hens are really settled down, and no eggs have been pushed out. Make a mark on each nest-box that has chipped eggs in it, and take a look at these in the evening, as shells may want removing or young may get jammed and suffocated in an empty shell, or you may have two or three chipped at the same time, possibly allowing you to put those that are already unhatched under one hen, the chipped under another, and unchipped under a third. If the young are nicely dried, it is not advisable to leave a hatched brood overnight in a nest-box, as room is limited, and some are apt to get squashed. The cheeping of the ducklings, too, disturbs the hens on each side that are perhaps not ready yet to hatch. If eggs are well chipped in the morning at the usual time for feeding the broodies, it is better not to remove the hen, for eggs are most easily broken at this stage, and no good hen will wish to feed with the eggs chipping under her.

BUYING BIRDS, AND THEIR TREATMENT
ON ARRIVAL

WHEN buying waterfowl, stipulate that they are hand-reared, for wild caught ones, that is the females, will rarely breed, though males will do so. It is not always easy to know a wild caught from a hand-reared bird, for if they have been kept a long time in close confinement where they are continually seeing people, they get fairly tame, but when put on natural ponds they very quickly revert to the wild. Examine the wing where pinioned. In hand-reared birds the cut pinion joint will be concealed by the growth of feathers, as the pinioning was done when the bird was young. In wild caught ones the cut will look bare and be much more easily seen, as it was done when the birds were adult. On the water, even if fairly tame, wild caught birds sit noticeably more erect and alert, and look more suspicious of danger. Birds bought unpinioned will be wild caught or ones that have been duck reared, probably surplus from some private collection, as I have never yet heard of any game breeder who did not pinion all his birds when young. Purchasers should always ask for and expect a guarantee with each bird, of sex, age, health, and whether hand-reared or not. Having bought the birds, it is advisable not to release them straight away, as the pond or lake may be large, and they may then have difficulty in finding the feeding place. Also, some varieties, Teal of all kinds more than other species, are very nervous, and are likely to hide if cover abounds and simply starve to death. The best

LORD GREY OF FALLODEN DEMONSTRATING TAMENESS OF HIS WATERFOWL

plan is to fence in a small portion of land and water close to and adjoining their future feeding place, where they can see the other duck coming and going from the food, thereby getting acquainted with them through the netting, and get a view of what will be their future home. It should be ascertained when buying the birds what food they have been accustomed to, and the same given to them for a few days. And if they are intended to be fed on different food in the future, this, too, can be added and the birds gradually accustomed to it. About a week to ten days should settle them into the food and get them acquainted with the surroundings, and they can then be released. Do not catch them, but leave an opening for them to swim out. This method of keeping them in a small pen has also the advantage of enabling one to find out whether any are in a poor condition or have been damaged in any way during transit, whereas on a large pond the condition of the birds would not be so easily seen. If the birds have been bought abroad and imported from some distance, great care must be taken of them, for they are sure to be in poor condition, and their plumage will be all dried up and broken and usually completely lacking in oil. If they should arrive in cold weather, put them into a shed for a time with a large dish—a shallow tin bath does well— in which they can wash. The water will soon get filthy and will have to be changed three or four times or more the first day. And after they have washed for a short time, it should be removed, so that the birds get time to dry and clean themselves between baths. Plenty of straw on the floor is soft for the birds'

feet, and drains off the water that is spilt or runs off
the birds' feathers. At the first washing it is as well
to remain with the birds to see that none stay in the
water too long and get drowned, which is quite likely
to happen if they are completely without oil, and also
to see they do not get chilled. Very nourishing food
should be given them, such as biscuit meal and some
chopped meat, worms and lettuce with the grain. On
board ship they may have been well fed, but are
almost sure to have gone short of green food and grit,
so do not forget the latter. Drinking water must be
given in dishes they cannot wash in, so that it is kept
fresh and clean. It will be easily seen when the oil is
beginning to function again, and when they are more
water-resisting they can be put outside in a pen with
plenty of shelter and a small pond until fit enough to
go to their permanent quarters. If unpinioned, no
attempt at pinioning must be made for a month, the
shock would be too much for them in a weakened state.
If several different species are obtained at the same
time, it may be noticed that some varieties are bullies,
and these must be kept separately. The best season
for purchasing waterfowl in this country is the autumn
and early winter, as it allows new arrivals time to get
to know the ground thoroughly, settled down with
any kind that already may be on the water, and get
paired off before the spring. With birds bought from
abroad, especially if from a warmer climate, it is very
desirable that they should arrive here in the spring,
as that gives them time to acclimatize before winter.

YOUNG CAROLINA AND ROSYBILLED POCHARD IN MIDDLE

FEEDING AND MANAGEMENT OF
ADULT DUCK

THE feeding of adult waterfowl is both a simple and inexpensive matter. The majority of surface feeding duck when kept on waters where they can get a little natural food, a few water snails, aquatic weeds or submerged grass will, with the addition of one handful of grain per bird per day, keep them fit. Wheat is best, but barley or oats will do quite well. Indian corn is too fattening, and the smaller duck cannot swallow it. During hard weather, when natural food is unavailable owing to ice and snow, some meat food will be found beneficial ; house scraps are appreciated. All food should be given under water, in water that is of about six to nine inches in depth, as this avoids waste from sparrows. Diving Duck, those that inhabit fresh water, and Shelduck, require more meat food to produce better fertility, but of course on waters where they can procure for themselves some fair quantity of natural food, this is not necessary. Grit is essential to the well-being of all waterfowl, and they will hunt for and find this for themselves, if any exists. But if there is no source of natural supply, it must be given. This can quite easily be done by placing a small heap of river or seashore gravel at the edge of the water. Waterfowl on the whole are very hardy, but what they do detest is cold driving winds with sleet or snow. Therefore some form of shelter from such winds should be provided in the shape of bushes growing low to the water's edge or reed-beds. The ideal quarters are those so sheltered but open to the south and sun, with grassy banks and slopes on which they can sit and sun themselves, one end of the water terminating in a bit of marsh where they can guzzle among the mud and slime to their heart's content, and fresh water, however small a run, passing through is a great asset. When the ground is all snow and frost severe, some heaps of cut reeds or straw laid along the edge of the water they have kept open will be a boon quickly taken possession of, to sit on and keep their feet warm. Eggs cannot be expected or got without the nesting cover being suitable ; nesting sites are to duck what houses are to us, indeed some birds seem more difficult to suit than we are. In early spring, duck are in the habit of wandering about in the evenings and early morning examining the ground with a view to future nesting accommodation. At this season great care should be taken to prevent the birds being frightened or disturbed, the result of the coming breeding season depends largely on this. No hard and fast rule can be laid down for the actual site and type of the nest of each species, because individuals vary in choice and their choice may be of a necessity limited, but the following are the usual or average. Surface feeding duck mostly prefer rather rank semi-dead grass, dry at the bottom and some little way from the water. These nests are often well concealed by being under an overhanging bunch of grass or in the centre of a tuft ; such sites are chosen by most of the Teal, Shoveler, Common Wigeon, American Wigeon and Bahama. Falcated, Gadwall and Chiloe Wigeon like rather higher cover such as nettles or low bushes, grass grown through. Some are tree nesting, and require boxes made for them. These should be one foot square with an entrance door four inches square. The box has to be up off the ground two to three feet, and a rough ladder made from the ground to the entrance. They like these boxes best when placed on a fallen tree or against a tree trunk or among bushes ; boxes set up in the open are seldom used. The ideal box is one made out of a hollow log. Carolina, Mandarin and Chestnut-breasted Teal use such boxes. The latter duck with us nearly always lay in the grass in a typical Teal site, but with others they use the boxes, One season a Chiloe Wigeon nested in one of the boxes. The Shelduck nest in holes or between clefts in rocks. so artificial rabbit holes have to be constructed for them, and the deeper they are the better they like them. The entrances to these holes should be concealed by a fallen tree or bush. Diving Duck such as Tufted and White-eyed Pochard like low reeds, rather dense and damp at bottom, but I have seen Tufted nesting in dry grass, in old nests of Coots and built-up nests in shallow water, so their choice is varied. Red-crested Pochard prefer when possible a mass of bent-over reeds or aquatic herbage under which they can make their rather clumsy nests. Rosy-billed Pochard only nest in high reed-beds, building a large nest up out of the water, so do Canvas-back and American Pochard. The Common Pochard's nest may be in any of the above mentioned sites. On a neighbour's place they nest on a small rhododendron-covered island, bare under the bushes of all cover. Here they build their nests of small sticks and dead rhododendron leaves ! Grey and Canada Geese like

NEST BOX SUITABLE FOR CAROLINA AND MANDARIN

best small islands grass covered, with here and there an odd bush growing near to which they usually place their nest, making little or no attempt at concealment. Magellan like more cover, often choosing a clump of reeds or low bushes. Geese seem to have a better instinct for choosing sites above possible flood levels than the duck. Cereopsis are weird in their choice of homes. I have known them nest in an open shed and at another time on a pile of stones. After the various birds, duck and geese, have once chosen their nesting sites, they, or a site in the near vicinity, will be used year after year, though other places may look to us as good or better.

EGGS AND FERTILITY

THE best results are obtained from the eggs of one's own birds. A quick cut to increasing stock might be expected from bought eggs or getting eggs from the nests of wild birds. In practice neither of these ways is really a success. If eggs are purchased, unless of some of the commonest species such as Wild Duck, they do not give a high percentage of young, for the eggs of many species of duck are bad travellers. Indeed some, Eider for example, will hardly travel at all ; twenty-five miles aboard a train and the vibration will. ruin them. The eggs of Wild Geese do travel better, but not for long distances. Only this year we got a number of Pink-footed Geese eggs from Iceland, and none hatched, though the eggs when broken were fresh enough. If eggs are taken from wild birds' nests, unless one gets them oneself and puts them straight under a broody, it is never certain whether or not they may have been incubated for a day. Eggs that have been incubated for however short a period and then allowed to get chilled are useless. When taking eggs from the nest the greatest care must be taken not to disturb the grass and vegetation round the nest or whatever material is used to cover the eggs. Duck, when laying, cover their eggs with grass or whatever growth the nest may be situated among, when incubating, with down. No eggs should be removed until four have been laid as, previous to that number, if the nest is disturbed, the bird is very apt to desert. For each egg taken out, one should be replaced by a dummy egg. The dummy eggs must be as like the original egg removed as is possible, and it is a good plan to mark each dummy with an indelible pencil so it cannot be mistaken for one of the real eggs. Bantam eggs make a good substitute for the eggs of Carolina, Mandarin, Wigeon, Gadwall and Teal for example. And for Red-crested Pochard, Common Pintail and those that lay greenish eggs, those of the Common Wild Duck will do well, and these dummy eggs must be fresh. If rotten or addled the duck are apt to leave the nest. The geese are easier to please, as the eggs of domestic geese will do for all species. When the egg is taken from the nest, each should be dated, name of species written on it and where found, giving to each nest a name or number and where found, thus enabling one to mix up the eggs that do take the same period of incubation under one broody ; or to keep them separate and thereby knowing individual broods, so that unrelated lots can be put together for breeding. Another advantage is that if birds are in separate pens or on separate ponds, one can find out which produces the best results. Clutches vary in numbers, an all-over average being from seven to twelve. Once I found the nest of a wild Tufted Duck with forty-four fresh eggs in it. A veritable flock of duck must have got to work over that nest! Geese usually lay four and up to seven, but the latter number is not common. Fertility of eggs depends on a number of causes, namely, the health and age of the birds, whether related or not, feeding, and privacy. Among Wild Duck and their allied species such as Yellowbills, Dusky or Spotted-billed Duck, fertility is and should be high, and so also with the Pintails and Wigeon. But lower rates of fertility are got with some of the rarer Teal and the Carolina. With the Diving Duck it is largely a matter of food, for without some meat form of diet fertility is poor. Natural food—water snails, tadpoles, worms and water-weeds—all tend to an increase of fertility. And if kept on ponds where they cannot get for themselves any of this natural food, as they would not do, for example, if kept on concrete ponds, some form of animal food should be added in the spring, as grain alone is not conducive to fertility. Over-crowding can lower the percentage of fertility to a great extent, owing to some drake or drakes being of more amorous and quarrelsome natures than their fellow drakes, causing fights to ensue. And though, if even not of a serious nature, the vanquished are often driven off the water, along with their wives, to hide and mope on the land, no fertility can be expected

BLUE SNOW GEESE (FOUR WEEKS OLD)

from these pairs. Sometimes too over-zealous drakes will commence to chase their ducks before these are ready to breed, even killing them—Gadwall and Common Wigeon are sometimes bad offenders in this respect. When drakes behave like this it can only be stopped by removing the drakes, or the offender can be caught up and put on to a new pond with his duck. The drake then is likely to behave himself at least for some little time, as he is timid of his new quarters and strange companions. I have often had pairs so moved breed quite successfully in a week or two. If the waters have plenty of small bays, streams or inlets, or better still, are comprised of a series of small ponds, these difficulties seldom occur, for there is much more privacy for each pair, and the ducks can, if they wish, hide more easily from the drakes, so more eggs and higher fertility will result.

MANAGEMENT OF YOUNG

THE most critical time in the life of young waterfowl is the first four to five days, mainly due to the fact that some kinds are difficult to start feeding. If the days are cold and wet when they hatch, they are not only liable to get chilled, but are far harder to get to feed than when the weather is warm and sunny. Chill is the one thing that must be avoided, also wet, for being brought up under a hen, they lack the oiling they would otherwise get by nestling into their mother's feathers. Carolina are susceptible to cold more than other duck when small. They should be kept in a coop which has a wire run, and not allowed on to water until about three weeks old, and then only if the weather is fine and sunny. To avoid their getting wet, and to teach them how to feed, use small tins about one and a half inches deep and eight inches across. Into these tins is exactly fitted an inverted turf of grass, water is poured into the tin enough to moisten the whole turf and no more, two or three small depressions are then made by pressing a finger into the mud of the turf, allowing the water to seep into these little hollows, into which is scattered a little finely chopped lettuce. Better still, duck weed, if it is obtainable, chopped worms, and a little fine biscuit-meal which has been previously soaked in cold water until evenly moist. The ducklings are then lifted up one by one and their bills dipped into the water and food. This is repeated several times, and some of the food is scattered on their backs, for they will always pick anything off one another's backs, and so they learn to feed. Any young duck that is a shy or backward feeder if treated like this should get a lease of life. After four or five days the meal can be increased, and so, too, the duckweed at the expense of the worms, as these are troublesome to get, and when rearing a large number of waterfowl are a great source of labour. As the young grow, the size of the dish can be increased, and a fountain given for drinking water. The next step is on to a pond. This must be large enough to wash in, but not so large as to encourage them to remain on the water too long. I have given no exact number of days for these succeeding stages, as time is entirely governed by the weather conditions and

how the birds progress. In the north here we have
naturally to go more slowly and gradually through
the stages than one would have to do in a more
favoured climate. All surface-feeding duck will do
well on the above treatment except Shoveler, who
require moister food and more and a longer diet of
worms. Diving Duck are different, as they must
have swimming water—we only keep them off for
the first twenty-four hours. They will live if treated
like the surface-feeding duck up to a certain stage,
when they begin to go back and soon die. The water
they are on, however small it may be, and it is better
small, must be fresh. They eat more green food, also
more animal food, as in nature they would eat a lot
of fresh water snails, so worms or some animal food
must be given. I think, too, that the Diving Duck
eat more grit. Give a dish of grit to each coop, using
river gravel which has been passed through a fine sieve.
The Shelduck family like the surface-feeding duck, do
well if kept off the water for a week or two. They eat
enormous quantities of green food and, like the
Diving Duck, require some animal food. Eider and
other sea duck or partially sea duck should be fed for
the first month, anyway three weeks, on chopped
earth worms mixed with grit, and they eat a terrible
lot ! After that time a little biscuit meal can be added
to the worms with a little very finely chopped raw
liver, gradually giving more liver and dropping out
the worms and adding a little finely chopped lettuce.
All the time the water must be of the freshest, and they
can be allowed to swim after two or three days. Grit is
very necessary, and should be given after they are

NESTING COVER SUITABLE FOR SURFACE FEEDING DUCK

three weeks old in a shallow dish, with a small lump of rock salt placed on the grit. Previous to three weeks they get the grit mixed with the food. They do better when kept in small broods of about four or five. At two months of age the food consists of 2 ozs. of chopped liver to about five times that bulk of biscuit meal, soaked that is, and some chopped lettuce—this ration per bird. Birds of nervous dispositions like Common Teal, White-eyed Pochard, etc., do far better when each species is kept to itself until half grown anyway. For the first week or so, young duck require feeding from 6 a.m. to dusk, at about two hours interval, their water should be filled up and changed often. Shortly, if doing well, they should develop the most enormous appetites, and then meals can be reduced to five or six feeds, and as they grow so can the meals be reduced in frequency by one's own judgment of the birds appearance and appetites, but not below three meals until the birds have acquired their feathers completely, then two meals a day will suffice. Grain should be given at about two months old, and it should be soaked and mixed with the biscuit meal for a while to get them accustomed to it. Biscuit meal should contain a good percentage of sound meat and be of fine grade. I have tried many makers, and only found one really suitable, but obviously I cannot give the makers' name as it would look too much of an advertisement. A point worthy of remembering is that the bantams and hens on the rearing field should be fed with whatever food their brood may be getting, that is until they are well on, or deaths will occur from the young duck eating grain they cannot digest.

Young geese live entirely on grass, clover, buttercup and seeding grass heads, and are better without biscuit meal being given to them before they show the quills coming. It can be given after that, but if the grazing is to their taste they will not eat much else. They want plenty of grit and fresh clean water, as they do not seem to care about going on to water when little. They can quite safely be given access to it, that is if they are with hens. If with their mothers, she will take them on to the water. All young water-fowl must be able to get shade from hot sun and shelter from cold winds. A rather good plan is to have some movable hurdles made of reeds or straw, as these will answer both purposes. Natural shelter such as bushes growing on the rearing field is to be avoided, as they conceal vermin. Young waterfowl are fortunately not subject to many ills, if fed and housed correctly. Odd ones will get cramp from staying too long on cold water, others get cramp from being kept off water too long ! Blistered feet may be a trouble in hot weather, but they grow out of that. Sunstroke if caught in time can be got over by immersing the sufferers head in water and then keeping it in the shade for some time. One of the worst troubles is, if the weather is cold and wet, the birds are kept off the water, in consequence not being able to wash. They get into a dirty and sticky state from wiping their bills on their backs after feeding and from walking over each other in the coop. Then the only thing one can do is to clean them by hand, using a sponge and warm water, but taking care not to wet them too much. A more dread ill is an eye one, which can be noticed by a white film appearing on the eyelid and often a frothy substance at the corners of the eye. This I believe to be infectious, and birds so infected are better destroyed unless very valuable. Repeated washings of the affected parts with warm water and an eye lotion, as supplied by chemists specialising in the ailments of poultry, will, if taken in time, cause a cure. Feather-eating mite is a perfect curse, noticeable first by the duck affected not going on to water. If picked up and examined, it will be seen that the feathers on the breast have a dead and broken appearance. The birds are not affected until feathers first begin to grow. Treatment being—with a fine pair of pincers or by hand pull out each of the broken feathers and a small area round the affected part of the sound-looking feathers as well, rub an ointment which is sold for this purpose well into and round the bare parts, keep the bird off water for an hour or so (not that it will want to go on the water, but the grass may be wet), and repeat treatment until you can see the new feathers beginning to grow. Some deaths are likely after young birds have been put out on to large ponds, chill and change of food being the usual cause, often developing into consumption. If a young bird so put out looks poorly and it can be caught up, a return to smaller and more sheltered quarters and back to richer food will often pull it through.

GROUP OF YOUNG DUCK IN FIRST WINTER FEATHER

IN THE REARING FIELD

THE most suitable rearing field is one sheltered from north and east winds, sloping slightly to the south and bounded on one side by water. The ground should be divided up into as many pens as are likely to be needed, each with a water frontage. Two feet high fences to divide by are quite high enough, with a higher fence round the whole ground. Half-inch netting is necessary at the bottom for the first two feet, and one-inch for the next two. Iron pins fine enough to pass easily through the half-inch mesh of about three feet in length are a great convenience, as with some lengths of half-inch netting the pens can be divided and sub-divided when the broods want smaller spaces or to take in whatever size of water is needed at the time ; these are then movable small pens within the larger or fixed pen. The advantage of these will soon be found in practice. A small, rather deep basket with lid is very useful for carrying down young from the sitting boxes to the rearing field. When selecting a site for a coop, see that the ground underneath it fits all its edges and is flat, or young duck will escape. It is a good plan for the first few days to put a sack under the coop or rather that portion of the coop used by the hen. The sack must be folded to fit the coop. If it projects out and beyond the coop, rain will wet the exposed portions and be absorbed through and all over the sack. Turn the sack each day, and when it gets dirty use a clean one. If coops have sacks under them it prevents the bantam from scraping, which, after sitting, she will be sure to want to do, and also keeps the young ducklings nice and warm and dry, and it prevents the ground from getting as fouled as it would otherwise do. Do not let young broods out straight away into the pen, but surround the coop with a small strip of netting, using the pins already mentioned to fix it with. These temporary small runs can be whatever size is wished. We use seven-yard strips in a circle with coop in centre, and use larger surrounds as required. Young Carolina and Mandarin Ducks, for a few days after hatching, are great climbers, so netting must be set up perpendicular, and when it joins the edges must not be rough, otherwise it will enable them to get a grip when trying to climb. Needless to say, the netting must fit tightly to the ground, for some newly hatched young are very nervous, and will find the smallest of hollows under the netting, and if they escape often go to the wrong coop and get killed, or go to no coop and get chilled and die, which they are liable to do on a cold day in a remarkably short time. Ducks grow fast, and after a few days they will have got to know their own foster-mother and coop ; also, they will have seen their neighbouring broods through the wire and got to know them without actually mixing with them. Now suppose we have six broods all fairly newly hatched together in one of the permanent divisions or smaller pens of the whole enclosure. Soon they will have outgrown the need of the seven yard strip of netting enclosure and want more space. If they were all let out at once or even two broods, a general mix-up might incur, so this is how to proceed. The netting is removed from one coop, the most advanced brood, and the day that it is taken away that coop is not moved ; otherwise all coops are moved every morning on fine days and if wet, often in the evening as well, for they get much dirtier inside on wet days. The reason in not moving the coop the day the netting is taken away, is that the young will find their own coop more readily. They will run all round the other coops and possibly try to get into them, but cannot do so, and will then go back to their own coop, where their water and food will be. In one or two days the next brood can be let out, and so on with the lot, until all are out. Treated like this, broods are not liable to mix if of different species. Of course, if all are of one kind, then they will mix up, but then all that is required, if they are of the same age, is to divide them up at night equally. Personally I prefer, when possible, different species together, for if a shower of heavy rain comes, each brood will go to its own coop, but if all of the same kind they are liable to all go into one coop, which gets into a horrible mess, and they are not sufficiently brooded. Seasons vary, but at our busiest time we usually have about ten broods in each pen. All must be shut up at night and each lot into its own coop, so it behoves one to remember the number and kinds in each coop. When young are three-quarter feathered, coops can be dispensed with, doing this to three or four broods or more in one day. This will avoid quarrelling for, if a new brood is added some days later, they may come in for a bit of bullying at night when all are shut up in the one general house, which is the stage they have now reached. They can be shut up so until large enough to go out on to the

SIX CAROLINA AND ONE MANDARIN (ALL TWO WEEKS OLD)

open ponds. The type of house used is long and low, with a wire netting floor deeply bedded with straw, which is renewed frequently. The house has one end open only, and that covered with half-inch netting, the whole being rat-proof. By the time they are old enough to go to the ponds they would be able to fly, but we always pinion them of course at three or four days old. If the risk of some straying and getting lost or shot in order to have birds flying about full-winged is considered worth while, a partial way out of the difficulty, whereby some birds can be kept full-winged, is to pinion only the females. The drakes then will not leave if there are sufficient ducks to go round. But by this method birds cannot be pinioned until their sex is known, necessitating clipping off the flight feathers up to that time. Duck pinion most easily at three or four days old. So little do they feel it that when pinioning a brood it is quite customary to have the bird that has first been done feeding before one has finished the last, and it only takes a few minutes to do the lot.

THE DUCK AS THE MOTHER

THIS obviously saves much work and trouble, but has disadvantages. Young so hatched are open to all attacks by vermin, for when they are little they must come ashore sometimes and, on the water, pike take a heavy toll. Eggs are eaten—rats, stoats, rooks, crows and even waterhens are not free of guilt in this respect. Often a bad-tempered old drake will kill young duck. Those that are reared must be caught and pinioned, no easy matter on even quite a small pond. If reared by the duck and left full-winged, I find them far more liable to wander than would hand-reared birds that were full-winged. Strange though it may sound, it has not been my experience that duck reared by their own mothers breed nearly as well as those reared by hand—the reason being that they pair with members of the same family often, and fertility is then very poor. Hand-reared, it is easy to pair up unrelated pairs. If it is decided to allow the ducks to rear their own young, vermin must be killed out as much as is humanly possible, especially so rats and rooks, these being, we find, the two worst offenders. But we have no foxes, or they would come very easily first, as they eat eggs, young and adults. At the time of hatching and onwards until the young can eat grain, some meal put at the usual feeding place will be found by most of the young, for the mothers will come for grain if a supply of natural food is not available. If it is the old birds will not bring their young to the feeding place, for it is her instinct to keep her brood away from any others. Some duck, such as Teal, are very nervous, and it is as well to have one or two extra feeding places, to give all a chance at this season. The food must be scattered in shallow water of such a depth that the young can readily find it. Shallow ponds, marshes and chains of little ponds, reed surrounded, are the ideal conditions for rearing broods on. Large ponds with plain grass edges produce little natural food and shelter, and whatever may be done in the way of artificial feeding, it is the natural food that counts when duck are left on their own, because they have none of the extra warmth of a coop or shelter at nights. Natural food for young duck for the first three or four weeks is mainly fly, with some water snails and aquatic weeds. I have even seen young Eider catching fly ! To produce then a plentiful supply of fly, the weather must be mild. During late frosts and cold wet weather, I have seen even the Common Wild Duck die off in large numbers when three-quarter grown, from starvation. These difficulties can be got over if each duck with her brood can be caught up just after they have hatched and put into a small enclosure of say 20 feet by 20 feet with a tiny pond in the middle with a little flow of water passing through, the merest trickle will suffice. In such conditions they can be protected from vermin, fed properly, pinioned easily, and a shelter provided for them. Treated so, few ducks will fail to rear their brood.

I have often heard it said that duck who nest in trees are in the habit of carrying their young down. On occasion they may do so, not always, as once I saw a Wild Duck on the ground at the foot of a fir tree calling to her newly hatched young, who jumped down from the nest, landing all of a heap, but apparently none the worse. The nest was about 15 feet from the ground, being an owl's nest of the preceding year. Carolina will do likewise, only, being pinioned, the nest-boxes are at the most, 3 feet from the ground, so the jump is of no height. One year we had a full-winged Carolina Duck who nested in a hollow branch of an old beech, at least 30 feet up. Unfortunately,

LORD GREY FEEDING DUCK. Note tameness of Full-Winged Mandarin Drake on seat

no one saw the young until she had got them on to the pond. Lord Grey of Falloden informs me that young Carolina were observed on his place jumping from the nesting hole, 21 feet from the ground. A strong wind was blowing, and they were blown sloping to the ground, but in no case were the young injured. Watching a marsh through glasses I saw a Teal rise and fly away. Thinking it was a duck leaving her nest to feed, I marked the place she rose from, and on reaching it I found nine young Teal of about three or four days old sitting in a compact bunch. As there was no apparent danger, and the young were not hidden, the conclusion one could come to was that the mother had left her young with instructions not to move, while she herself went off for a short while in search of food or grit.

Two drakes will occasionally pair, or rather form a strong friendship, keeping together throughout the breeding season, allowing no female near them. Two Common Wigeon drakes did this one season with us, and at another time two drake Carolina behaved in like manner. Catching these birds up in the autumn and putting them in separate pens, the following spring they behaved normally and paired with duck of their own kind. Two duck will sometimes attach themselves to one drake, despite the attentions of the drake who should have otherwise claimed one of the ducks. A goose or gander once paired to a species other than its own, will always try to pair with that kind, even if it cannot see the bird, but is within hearing of it. They will spend most of the day calling back and forwards to each other, taking no notice of their own species in the pen with them. Geese breed in their third year, but it is quite common to get clear eggs in the first nest, though one should expect fertile ones afterwards. I had an adult Blue Snow Goose whose gander died, and the only other available gander was only one year old. The goose laid six eggs. One hatched, which is the only case I have known of a young gander at that age fertilizing any eggs. When a drake or gander becomes quarrelsome, if possible it should be caught and put into a strange pen with birds larger or stronger than itself. Often this will cure it of bad habits, anyway for some time. Hybrids on the whole are comparatively rare, but some species hybridise more readily than others, such as the Mallard group. And if it is wished to keep on the same water several varieties of this family, each pair should be kept separately until properly paired before putting out all together. Once paired, they usually keep so from year to year. Wild caught ducks will very rarely ever breed, but the drakes will do so, and are splendid for a complete change of blood. Being wild caught, they are naturally wild, so if turned out after arrival on to large waters or even comparatively small ponds with cover round, are very apt to hide and possibly starve to death. Therefore they should be kept in a small pen with no cover and only a small pond, together with the hand-reared duck they are to pair with, for some weeks or until they at least become moderately tame. Those ducks that show some black spots at base of bill, show age by a corresponding increase of size and number of these spots. Duck with white on forehead and round the eye show age by increase in size of these

WINTER SHELTER

white portions. Geese, that is Grey Geese, show age
by an increase of black feather on the breast or white
on the forehead. The useful laying life of the average
duck is to about ten or eleven years, geese living very
much longer. We have some now laying fertile eggs
that are nearer thirty than twenty, and I have known
them live over half a century. Dates of laying of
individual birds vary very little from year to year,
provided the weather is not abnormal and the sites
of the nests are usually in close proximity to those of
the preceding year, if they have been allowed to hatch
out one or two young. Waterfowl are capable of
looking after themselves most of the year, the dangerous
period being of course during laying and sitting. Even
then vermin, unless foxes can get them, take a very
small toll. The eggs are more likely to suffer from rats,
stoats, crows, rooks and even waterhens. Duck
cover the eggs whenever they leave the nest, but if
disturbed will leave them uncovered. Young duck
unfortunately have some enemies they simply cannot
escape, if these abound, namely pike when on the
water, and rats on land. The domestic or Mute Swans,
if they should have a nest, will kill all young duck
who may come near, as they will often do without a
nest as an excuse. All hand-reared duck offered for
sale in this country are pinioned, so if the water they
are to be kept on is suitable to waterfowl, they will
stay, and there is little need for fencing. But if it
is considered necessary, a fence 4 feet high will be
sufficient to keep them in. Where foxes are preserved,
a fence 6½ feet high with the bottom 3½ feet, of 1-inch
mesh with 6 inches sunk in the ground and the remain-

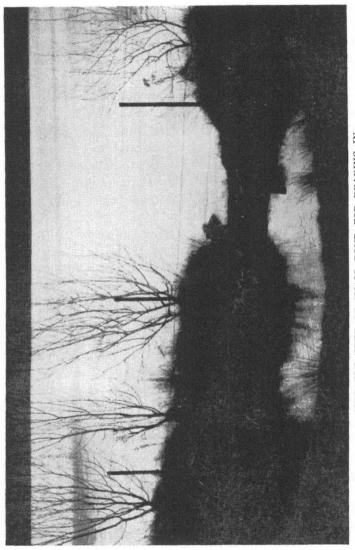

SUITABLE TYPE OF SMALL POND FOR REARING IN

ing 3 feet of 3-inch mesh, will suffice. When erecting the fence, care must be taken that the immediate ground outside the fence is not sloping up, as that would enable a fox to jump more easily over, and no branches or trunks of trees must overhang the fence. The fence should not be placed close to the water, for many species of duck like to nest at some considerable distance from the water, and like to wander about at dusk and dawn over the grass in search of slugs and worms. Any enclosure for waterfowl can be of a permanent nature, for duck do not foul the ground like poultry or pheasants.

Wild Duck of several kinds visit our ponds, and to keep their numbers down they have to be shot. The hand-reared birds are so confident in man that not only do they take no notice of shots fired close to them, but show no fear even when a shot bird may fall among them. If wild birds are allowed to remain undisturbed, they, too, become very tame. One year a Wild Greylag took up his quarters on a pond with the hand-reared Greylag and, after a few days became as tame as them, but when migration time came round, left on May 5th. Since writing the above the Greylag has returned, on January 10th, and straight away swam up to be fed.